Manufacturing Process Validation

Unlock the secrets to optimizing manufacturing processes with this essential guide to process validation. Designed for both industry professionals and academics, this book bridges the gap between theory and practice, offering a clear roadmap to enhance quality and productivity. Discover core concepts, step-by-step implementation strategies, and powerful problem-solution tools. Explore the real-world case studies that showcase how process validation transforms production lines, from data collection to actionable improvements. Whether aiming to streamline operations or achieve operational excellence, this book provides the knowledge and practical insights to drive continuous improvement.

Manufacturing Process Validation: Concepts, Tools, and Industrial Applications discusses the importance of manufacturing process validation in addressing quality issues. It explores the primary tools utilized in implementing process validation within industrial settings and features two case studies demonstrating how validation can enhance production processes. Process validation is crucial in guaranteeing a process's quality and efficiency. The book goes on to emphasize the significance of collecting and evaluating data from the design phase through production to establish a process's quality and reproducibility.

Designed for professionals, this resource serves as a crucial link between theoretical concepts and real-world applications. It provides a comprehensive guide for successfully implementing process validation, offering a clear and detailed roadmap for achieving optimal results.

Manufacturing Process Validation

Concepts, Tools, and Industrial Applications

Arturo Realyvásquez-Vargas, Jorge Luis García Alcaraz, Suchismita Satapathy and José Roberto Díaz-Reza

CRC Press
Taylor & Francis Group
Boca Raton London New York

CRC Press is an imprint of the
Taylor & Francis Group, an **informa** business

First edition published 2026
by CRC Press
2385 NW Executive Center Drive, Suite 320, Boca Raton FL 33431

and by CRC Press
4 Park Square, Milton Park, Abingdon, Oxon, OX14 4RN

CRC Press is an imprint of Taylor & Francis Group, LLC

ISBN: 978-1-041-09609-2 (hbk)
ISBN: 978-1-041-09610-8 (pbk)
ISBN: 978-1-003-65096-6 (ebk)

DOI: 10.1201/9781003650966

Typeset in Times
by Apex CoVantage, LLC

Contents

Preface

Today, companies are under pressure to meet increasing global competition, stringent product quality standards, and environmental regulations. In this context, process validation is gaining relevance. This concept, which involves confirming through documented, objective evidence that the processes used in manufacturing a product or providing a service meet the specified requirements, is a key tool for improving efficiency in industries by reducing overall energy and resource consumption.

In this context, process validation is crucial for ensuring the quality and reproducibility of a process. It involves the collection and evaluation of data from the design phase to production, to establish the quality and reproducibility of a process. This includes identifying and documenting the critical parameters of the manufacturing process, as well as the requirements of raw materials. By doing so, companies can be confident that their products or services meet the required specifications, even if defects are not evident until the product is in use or the service is delivered.

Currently, in the specific case of the manufacturing industry, quality inspection has become a critical aspect, where products are evaluated and determined to be either acceptable or rejected. Therefore, process validation is a fundamental requirement in this sector, especially in production environments requiring a high degree of control and precision, such as those involved in manufacturing pharmaceuticals and medical devices. Through process validation, manufacturing companies can demonstrate that a process can consistently produce a product that meets predefined specifications and quality requirements in a consistent and controlled manner.

Moreover, process validation plays a pivotal role in ensuring product quality and safety, particularly in the context of Good Manufacturing Practices (GMP). This regulatory requirement, established by agencies such as the U.S. Food and Drug Administration (FDA) and the European Medicines Agency, is underscored by ISO 13485. The ISO standard, which provides requirements for medical device quality management systems, highlights the importance of process validation in meeting predefined specifications and quality requirements.

The impact, importance, and need for manufacturing companies to validate their processes are so great that a search on Google Scholar with the word "manufacturing process validation" yields approximately 3,300,000 publications. Of these, 35,200 were from 2020 to the present (October 5, 2024). Given the importance of process validation in the industry, this book provides a conceptual framework, describes auxiliary tools used during its execution, and provides two case studies. The book is divided into four chapters.

Chapter 1 presents an overview of the validation process for manufacturing. It provides a definition and presents its evolution in the manufacturing field, as well as the procedure, barriers, critical factors, and benefits of its implementation. Finally, a bibliometric review of process validation is offered to understand the state of the art, prominent authors, journals, and research groups focused on this area.

Chapter 2 provides a detailed description of the different methods and auxiliary tools used for process validation. These include the 8D methodology, operation flow diagram, process flow diagram, Pareto diagram, Ishikawa diagram, analysis of variance (ANOVA), and design of experiments (DOE). This description includes the concepts, advantages, and disadvantages of each tool at the time of its application.

Chapter 3 presents a case study validating a sealing ring annealing process carried out in two plants of the same company. Initially, the process lasted 344.08 hours and produced 5,185 defective parts. After process validation, the process was reduced to 24.08 hours, defects decreased to 3,736, production increased, and the demand was met.

Chapter 4 presents the second case study, which validates the introduction of an ultrasonic sealer into the laparoscopy model production process. Therefore, its validation was required using IQ, OQ, PQ, and a Cpk greater than or equal to 1.33, as well as determining its optimum operating parameters. The equipment was validated using both the DOE methodology and an internal company methodology, resulting in a Cpk of 6.34.

About the Authors

Arturo Realyvásquez-Vargas is a full-time professor from the Department of Industrial Engineering at Tecnológico Nacional de Mexico/Instituto Tecnológico de Tijuana in Mexico. He received a master's degree in industrial engineering, a Ph.D. in Engineering Sciences from the Autonomous University of Ciudad Juarez in Mexico, and a Ph.D. in Innovation in Product Engineering and Industrial Process at the University of La Rioja (Spain). His main research areas are industrial process optimization, lean manufacturing, and ergonomics. He is an active member of the Society of Ergonomists of Mexico Civil Association (Sociedad de Ergonomistas de México, SEMAC) and the Academic Network of Optimization of Industrial Processes (Red de Optimización de Procesos Industriales, ROPRIN). Dr. Realyvásquez is a national researcher recognized by the National Council of Science & Technology of Mexico (CONACYT) as level I. Furthermore, Dr. Realyvásquez is an author/coauthor of around 20 papers published in journals indexed in the *Journal Citation Reports*. He has attended international conferences and congresses in Mexico and the United States of America.

Dr. Realyvásquez has supervised more than 20 bachelor theses and five master theses. In addition, Dr. Realyvásquez is the author of a book published by Springer about ergonomics and has edited two ergonomics-related IGI Global books.

Jorge Luis García Alcaraz is a full-time researcher at the Industrial Engineering and Manufacturing Department at the Autonomous University of Ciudad Juarez in Mexico. He received an MSc in Industrial Engineering from the Colima Technological Institute (Mexico), a Ph.D. in Industrial Engineering Sciences from Ciudad Juárez Technological Institute (Mexico), a Ph.D. in Innovation in Product Engineering and Industrial Process from the University of La Rioja (Spain), a Ph.D. in Sciences and Industrial Technologies from the Public University of Navarre (Spain), a Ph.D. in Mechanical Engineering from the University of Zaragoza (Spain), and a Postdoc in Manufacturing Process from University of La Rioja (Spain). His main research areas are related to the multicriteria decision-making process and techniques applied to lean manufacturing, production processes, and supply chain modeling. He is a founding member of the Mexican Society of Operations Research and an active member of the Mexican Academy of Industrial Engineering. Dr. García is a national researcher, recognized by the National Council of Science and Technology of Mexico (CONACYT) at Level III. In addition, Dr. García is an author or coauthor of approximately 215 documents indexed in Scopus (including papers, books, and conference proceedings) and has attended over 150 international conferences and congresses worldwide.

Moreover, Dr. García has supervised over 50 bachelor theses, 18 master theses, and 13 doctoral theses. He has 13 international patents and one industrial design in Mexico and Spain. Dr. García is also the author of several books published by international publishers such as Springer and IGI Global, all of which relate to lean manufacturing, its tools and techniques, kaizen, supply chain management, and just-in-time philosophies. Dr. García won the Chihuahua State Science and Technology Award in 2015 and is a member of several doctoral committees at universities in Mexico, Colombia, and Spain.

Suchismita Satapathy is an associate professor in the School of Mechanical Sciences, KIIT University, Bhubaneswar, India. She has published over 130 articles in national and international journals and conferences. She has also published many books and e-books for academic and research purposes. Her areas of interest include production operation management, operation research, acoustics, sustainability, and supply chain management. She has filed three Indian patents, two of which have been published. She has over 15 years of experience in teaching and research and has guided numerous Ph.D. students (two completed, four ongoing), M.Tech. students (20), and B.Tech. students (49). Dr. Satapathy has published books such as *Production and Operations Management (Stadium Press) and MCDM Methods for Waste Management* with CRC Press, Innovation, Technology, and Knowledge Management with Springer, and Soft Computing and Optimization Techniques for Sustainable Agriculture by De Gruyter. Many of her projects are ongoing.

José Roberto Díaz-Reza holds a master's degree in industrial engineering, with a specialty in quality, and a Ph.D. in innovation in product engineering and industrial processes from the University of La Rioja (Spain). In addition, he has a Ph.D. in advanced engineering sciences from the Autonomous University of Ciudad Juárez (Mexico). He has published a book about total productive maintenance with Springer and some articles in indexed journals related to lean manufacturing tools. Dr. Diaz-Reza has also participated in international congresses, and his areas of interest are optimizing industrial processes using lean manufacturing tools.

Process Validation *Concepts and Definitions* 1

1.1 INTRODUCTION

Manufacturing process validation is a continuous, rigorous, and detailed process that confirms production processes can meet the quality requirements established for a specific product (Anju and Pandey 2017). This validation represents a critical element in ensuring that manufacturing processes are consistent and predictable, thereby ensuring the uniform quality of the final product (Guo et al. 2018). This quality is crucial for product safety and efficacy, especially in the pharmaceutical and medical device industries (Sastri 2014). This involves a thorough evaluation and documentation of each step in the manufacturing process, from raw materials to the final product, as well as equipment calibration and maintenance, and personnel training (Ryabchik et al. 2019). This ensures that each stage meets the quality standards and that the regulations and requirements agree with the customer's needs (Mathiyazhagan et al. 2022).

Manufacturing process validation involves identifying critical parameters, performing control tests, and collecting and analyzing data to monitor process performance over time (Weiss et al. 2015). Additionally, it is crucial to periodically review standard operating procedures to ensure that they remain effective and meet the required quality parameters, train personnel, and implement corrective actions in the case of deviations (HEFLO 2024). Regular training of personnel in these operations and the implementation of corrective actions in the case of deviations are fundamental aspects of manufacturing process validation (Pharma Times Latest Drug Updates 2023).

DOI: 10.1201/9781003650966-1

In the specific case of the pharmaceutical and medical device industries, where product safety and efficacy are of paramount importance, rigorous process validation is essential. This ensures that products comply with established regulations and requirements and maintain consistent quality, which is critical for protecting patients' health and well-being (Anju and Pandey 2017). Therefore, process validation is not only a regulatory requirement but also an essential practice to ensure the integrity and reliability of manufactured products (Jatto and Okhamafe 2002). In addition, moral standards can influence perceptions of individual and collective responsibility toward the environment, both individually and collectively; therefore, it is necessary to validate processes and avoid pollutants (Partala and Saari 2015).

1.2 ORIGINS OF MANUFACTURING PROCESS VALIDATION

The study of manufacturing process validation is rooted in the need to ensure quality and safety of the manufactured products (Sandhya et al. 2015). Since the Industrial Revolution, companies have sought ways to ensure that their manufacturing processes are consistent, reliable, and capable of producing products that meet the quality standards (Alhuraish et al. 2017). Process validation has become a fundamental part of the manufacturing industry, ensuring that manufactured products are safe for their intended use and meet government regulations and customer requirements (Jatto and Okhamafe 2002). This focus on process validation has led to the development of sophisticated techniques and methodologies to continuously evaluate and improve manufacturing processes, which, in turn, has contributed to the evolution and growth of the modern manufacturing industry (Anderson et al. 2011).

Manufacturing process validation has evolved and undergone several key stages. The following is a chronological summary of the main stages of process validation development.

- Beginning of standardization (1900s): During the early 1900s, companies began to recognize the importance of standardizing their manufacturing processes (Pazhayattil et al. 2018). This involved establishing specific procedures and requirements to ensure the consistency and quality of the manufactured products (QMPG 2017).
- Statistical quality control approach (1920s): In the 1920s, statistical quality control—a methodology that involved using statistical techniques to monitor and control manufacturing processes—was

developed and popularized (de Toledo et al. 2017). This allows companies to identify and eliminate sources of variability in their processes, thereby improving the quality of manufactured products (De Mast and Does 2006).

- Focus on total quality and quality control (1940s–1950s): In the 1940s and 1950s, there was a shift in focus toward total quality and quality control in the manufacturing industry (Guo et al. 2018). During this period, more comprehensive methodologies and approaches were developed to address quality in all aspects of the manufacturing process (Mathiyazhagan et al. 2022). This included adopting quality control practices throughout the product life cycle, from the design and development stages to production and after-sales services (Encyclopedia.com 2018).
- Risk-based process validation and compliance (1980s–1990s): During the 1980s and 1990s, a risk-based process validation and regulatory compliance approach emerged. This approach involves identifying the potential risks associated with manufacturing processes and developing strategies to mitigate and control these risks (Kneat 2024).
- Process validation as part of the product life cycle (2000s to present): In the 2000s and up to the present, process validation has become an integral part of the product life cycle (Sandhya et al. 2015). This approach recognizes that process validation is not a one-time event but a continuous process that must be carried out throughout the product life cycle (Pazhayattil et al. 2018). This ensures that manufacturing processes are consistent and meet quality requirements at all stages, from manufacturing to after-sales services (Encyclopedia.com 2018). Thus, process validation has become a standard practice for ensuring the consistency and quality of manufactured products (Sandhya et al. 2015).

1.3 MANUFACTURING PROCESS VALIDATION INDICATORS

Manufacturing process validation can be evaluated using several key indicators. Critical indicators include process capability, process variability, equipment capability, throughput rate, and product quality (GMP7.COM 2019). These indicators are crucial to ensure that the manufacturing process is validated and that the products are consistently of high quality.

Each of these indicators is briefly described in detail.

1.3.1 Process Capability

Process capability refers to the ability of a process to produce products with specified tolerances (Yao et al. 2017). It is a critical indicator used to measure a manufacturing process's ability to produce products that meet the required specifications (Advance Electrical Design & Engineering Institute 2024). This is expressed by a process capability index, commonly referred to as *Cp*, which is calculated using Equation (1.1) (Walsh et al. 2018):

$$Cp = \frac{USL - LSL}{6\sigma} \tag{1.1}$$

where *USL* is the upper specification limit and *LSL* is the lower specification limit; in other words, they are the limits within which the product must comply. By contrast, σ represents the standard deviation of the process, which measures the inherent variability of the process (Avramova et al. 2024).

A *Cp* value greater than 1 indicates that the process variability is acceptable compared to the specifications, suggesting that the process can produce products within established limits (Walsh et al. 2018). However, a *Cp* value of less than 1 indicates that the process variability is too high compared to the specifications, meaning that the process could produce products outside the required limits (Avramova et al. 2024). Table 1.1 shows the interpretation of a manufacturing process's capability according to *Cp* index values.

TABLE 1.1 Interpretation of *Cp* index values

CP INDEX VALUE	*PROCESS CATEGORY*	*DECISION (IF THE PROCESS IS FOCUSED)*
$Cp < 0.67$	4	The process needs to be more adequate. Severe and immediate modifications are required
$0.67 < Cp < 1$	3	The process needs to be more adequate. It needs to be analyzed and modified to reach a satisfactory level of quality
$1 < Cp < 1.33$	2	The process is partially adequate but requires strict control
$Cp > 1.33$	1	The process is adequate
$Cp \geq 2$	World class	Quality is Six Sigma

Source: Adapted from Gutiérrez Pulido and de la Vara Salazar (2013)

The process capability has significant implications for the quality and consistency of manufactured products. A process with high capability is less likely to produce defective products, leading to fewer reworks, lower associated costs, and fewer dissatisfied customers (Purba et al. 2021). Therefore, continuous evaluation and improvement of the process capability are essential to ensure the quality and efficiency of manufacturing (Mathiyazhagan et al. 2022).

1.3.2 Process Variability

Process variability is another crucial aspect of manufacturing process validation, referring to the extent to which process results deviate from their expected average value (Lee et al. 2004). This variability can be quantified using the standard deviation of the process, represented by the Greek letter sigma (σ) (V. Wang et al. 2009). When variability is high, results are scattered and less predictable, resulting in a greater likelihood that products will not meet established specifications, which can lead to higher defects and increased costs associated with poor quality (Sadiq et al. 2021). On the other hand, low process variability leads to greater consistency in production, resulting in a lower rate of nonconforming products, thus providing compliance with customer specifications (Yao et al. 2017).

Process variability is expressed as the total variation, which is the sum of variations due to both common and special causes (Yao et al. 2017). Common causes are inherent to the process and affect all units produced, whereas special causes arise from specific circumstances and are not inherent to the process (Tonini et al. 2007). Different statistical tools can be used to measure it, such as control charts and ANOVA, to decompose process variability and understand its sources (Korzenowski et al. 2015).

To assess process variability, it is necessary to calculate the process capability index represented by *Cpk* in terms of variability. This index considers whether the process mean is centered in relation to the specification limits and quantifies the ability of the process to meet the required tolerances (Balamurali 2022). *Cpk* was calculated using Equation (1.2) (Balamurali 2022):

$$Cpk = Min\left[\frac{USL-\mu}{3\sigma}, \frac{\mu-LSL}{3\sigma}\right] \tag{1.2}$$

where μ denotes the process mean. A *Cpk* value greater than 1 indicates that the process is capable of producing products within specifications, considering both variability and the position of the mean relative to the specification limits (Walsh et al. 2018). Conversely, *Cpk* a value of less than 1 suggests that the process has difficulty meeting specifications owing to variability and/or the position of the process mean (Dudek-Burlikowska 2005). The *Cpk* index was always less than or equal to the *Cp* index. When the two are very close, the mean of the process is very close to the midpoint of the specifications; therefore, the potential and actual capacities are similar. When the process mean is out of the specification, the values of the *Cpk* index will be zero or negative (Gutiérrez Pulido and de la Vara Salazar 2013).

This indicates that process variability is critical in manufacturing process validation, as it directly influences product conformity and quality requirements (Zhao et al. 2017). Therefore, understanding and controlling process variability is critical for ensuring consistency and quality in manufacturing (Ryabchik et al. 2019).

1.3.3 Equipment Qualification

Equipment qualification consists of three main phases: installation qualification (IQ), operational qualification (OQ), and performance qualification (PQ) (Anju and Pandey 2017). It comprises three main phases: installation, operational, and PQ (Anju and Pandey 2017).

1.3.3.1 Installation Qualification

IQ focuses on verifying whether the equipment, systems, and devices used in the manufacturing process are installed correctly and operates according to the manufacturer's specifications to ensure quality and efficiency in product production (Mathiyazhagan et al. 2022). This includes reviewing equipment documentation, installation testing, and instrument calibration (Prism Group of Company 2024).

IQ is a comprehensive process that ensures the equipment and systems used in the manufacturing process are installed correctly and satisfies the manufacturer's specifications (Quality International 2024). This is accomplished through a detailed review of equipment documentation and instructions, identification of critical components, quality standard verification, and required instrument calibration (Lam 2004). One of IQ's critical aspects is installing installation tests that evaluate the functionality of the equipment and systems in their operating environment (CEA Instruments 2016). These tests ensure that components are connected correctly, interfaces function and integrate as intended, and control systems are configured appropriately (Software Testing Company 2022).

IQ is often expressed in the form of Equations 1.3–1.5, which represent the functionality, reliability, and efficiency of the equipment and systems, respectively (Y. Wang et al. 2012). These equations are critical for evaluating and documenting the conformance of a facility to the required standards and specifications (Adams 1919).

$$Funcionality\,of\,the\,equipment = \left(\frac{Number\,of\,functional\,components}{total\,number\,of\,components}\right) \times 100 \tag{1.3}$$

$$Reliability\,of\,the\,equipment = \left(\frac{Failure-free\,operation\,time}{Total\,operation\,time}\right) \times 100 \tag{1.4}$$

$$Eficiency \text{ of equipment} = \left(\frac{Real\,production}{Theoretical\,production}\right) \times 100 \tag{1.5}$$

Thus, Equations 1.3–1.5 provide a quantitative measure of the ability of equipment and systems to function properly, operate reliably, and produce results efficiently (De Ron and Rooda 2006). Evaluating these parameters helps objectively document the facility's conformance to the required standards and specifications, which is crucial for validating manufacturing processes (Teuchert 2015).

The value of IQ lies in its ability to establish a solid foundation for the reliable operation of equipment and systems, which in turn influences the integrity of the manufacturing process and the quality of the final product (Yao et al. 2017). Therefore, IQ significantly affects a facility's ability to comply with industry regulations and produce products that meet the established quality standards (Maishi MFG 2022).

1.3.3.2 Operation Qualification (OQ)

On the other hand, OQ is a fundamental phase in the validation process, in which the equipment or system is evaluated consistently and reliably under normal operating conditions (Sandhya et al. 2015). OQ is a critical phase after facility qualification, and is essential for demonstrating that the equipment operates properly under normal operating conditions (Anju and Pandey 2017). That is, OQ focuses on determining that the equipment operates within the established parameters and that operators are properly trained for its proper use (Tripathi et al. 2021). To calculate the OQ, specific equations and values are used to evaluate the equipment's operation, performance, and ability to satisfy the required technical specifications (Saravacos and Kostaropoulos 2016). These equations are fundamental to ensure that the equipment operates efficiently and predictably within established limits.

In the case of a manufacturing system, the OQ equations can include performance parameters such as production speed, equipment accuracy, and product uniformity (Bai et al. 2007). For example, production speed can be calculated using Equation (1.6) (Mathiyazhagan et al. 2022):

$$\textit{Equipment production speed} = \left(\frac{\textit{Units produced}}{\textit{Production time}}\right) \times 100 \tag{1.6}$$

The values obtained from the OQ are indicative of the reliability and consistency of the equipment or system under normal operating conditions, which have direct implications for the quality and integrity of the manufacturing process and, ultimately, for the quality of the final product (Mathiyazhagan et al. 2022). These values also enable the documentation and demonstration of compliance with quality standards and applicable regulations (Maishi MFG 2022). Thus, OQ equations are crucial tools for evaluating and documenting the performance and reliability of equipment or systems, which helps ensure quality and efficiency in the manufacturing process.

It is important to note that OQ focuses not only on evaluating the equipment itself but also on the training and competence of operators to use it properly (Leclair 2013). In this sense, equations and values associated with OQ can also include training indicators, such as accuracy in handling the equipment, understanding of operating procedures, and application of best practices in using the equipment (Dong et al. 2012).

1.3.3.3 Performance Qualification

Finally, PQ is an essential stage in industrial process validation because it focuses on demonstrating that the equipment or system can function effectively and reliably under normal operating conditions (Walsh et al. 2018). During the PQ, tests and measurements were conducted to determine the ability of the equipment to produce consistent and high-quality results. PQ encompasses several performance dimensions, including productivity, quality, and sustainability, and is often influenced by organizational practices, industry standards, and external economic conditions (PIA Automation 2024).

An important aspect of PQ is its reliance on performance feedback mechanisms, which are crucial for organizational adaptation and improvement. Organizations often compare their performance with that of a benchmark group of comparable companies to gauge their position in the industry. This benchmarking process is essential for understanding the relative performance and making strategic decisions that will enable them to improve (Mathiyazhagan et al. 2022). In addition, they enable compliance with quality standards and applicable national and international regulations, which are critical for process validation in industry and its certification (Maishi MFG 2022).

IQ, OQ, and PQ are critical stages in process validation because they ensure that the equipment (machinery and personnel) can operate effectively and reliably, meeting the established quality standards. Additionally, using specific equations and values during these stages is essential for evaluating and documenting equipment performance under normal conditions, which contributes to the integrity of the manufacturing process, quality assurance of the final product, and comparisons with competitors.

1.3.4 Rate of Return

The performance of a manufacturing process refers to its ability to produce high-quality products within the required specifications (Mathiyazhagan et al. 2022). The throughput rate, also known as the process yield rate, is a critical indicator that assesses the efficiency and effectiveness of the manufacturing process (Grozdić and Demko-Rihter 2023) and is calculated using Equation (1.7) (Realyvásquez-Vargas et al. 2021):

$$Yield = \left(1 - \frac{Defective\,units}{Total\,units}\right) \times 100 \tag{1.7}$$

The numerator of the equation represents the units of product that conform to specifications, that is, the total units produced minus the ratio of defective units to the total units produced. Note that the denominator includes all units produced, regardless of their conformance to the specifications.

A high yield rate indicates that the manufacturing process can generate a high proportion of conforming products related to a lower number of defective products and a reduced risk of quality deviations (Colledani et al. 2014). However, a low yield rate suggests that the process does not produce a significant proportion of conforming products, which can lead to higher quality costs and decreased customer satisfaction (Teplická et al. 2023).

This indicates that the yield rate is a key indicator for evaluating the effectiveness of the manufacturing process and its ability to meet the quality requirements (Bai et al. 2007). In addition, it allows the identification of areas of opportunity to improve efficiency and reduce the number of nonconforming products, which in turn positively impacts the firm's profitability and reputation (Tripathi et al. 2021). Continuous evaluation and monitoring of the process yield rate are essential to ensure efficiency and quality in manufacturing production, and to identify opportunities for improvement that boost competitiveness and customer satisfaction when deviations are observed (Alhuraish et al. 2017).

1.3.5 Product Quality

Product quality is critical in manufacturing because it directly influences customer satisfaction, company reputation, and profitability (Mathiyazhagan et al. 2022). Product quality is defined by its ability to meet customer requirements and expectations as well as its consistency and reliability over time (Lina 2022). Product quality can be measured using various parameters, such as conformance to specifications, durability, reliability, performance, customer satisfaction, and customer complaints (Grace et al. 2021). In addition, product quality is closely related to the variability of the manufacturing process, equipment capability, and throughput rate, as described earlier, as these aspects directly influence the characteristics and performance of the final product (Tripathi et al. 2021).

Companies constantly seek to improve the quality of their products, as this allows them to differentiate themselves in the market, gain customer confidence, and increase customer loyalty (Robinson and Etherington 2006). Implementing continuous improvement strategies, using quality control tools such as statistical process control, and committing to operational excellence are key elements in ensuring and elevating product quality (Modgil and Sharma 2016). It is critical to understand that product quality is not limited to the absence of defects but involves delivering a product that meets customer needs and expectations consistently and on time (Kenyon and Sen 2015). Some authors see product quality as a reflection of the company's commitment to excellence, innovation, and customer satisfaction, and as a determining factor in competitiveness and long-term success (Psomas and Jaca 2016).

It can be concluded that product quality is a fundamental attribute that impacts customer perception, company reputation, and business profitability; therefore, it must always be monitored. Understanding and managing product quality requires a holistic approach that considers process variability, equipment capability, rate of return, and continuous commitment to improvement and operational excellence.

1.4 PROCEDURE FOR THE IMPLEMENTATION OF PROCESS VALIDATION

The procedure for implementing process validation consists of a series of steps that must be developed sequentially to obtain better results from the company's strategic plans and programs. The steps are presented as follows.

1. Defining the objectives of process validation: It is important to clearly state what you are seeking to achieve with process validation and what you are doing it for (Anju and Pandey 2017). This could include ensuring product quality and safety, complying with regulatory requirements, and improving operational efficiency (Sandhya et al. 2015).
2. Identifying processes to be validated: To ensure product quality and safety, critical processes must be identified and validated (Pazhayattil et al. 2018). This is done by identifying the parameters and analyzing those that fail to comply with them.
3. Develop a validation plan: A detailed plan should be developed that includes the resources needed to conduct process validation (GMP7.COM 2019). The plan should include the identification of critical process parameters, the development of validation protocols, and the timing and assignment of responsibilities to the work team members; that is, it is a detailed schedule for the execution of the tests that include the names of responsible parties, structures, times, and ways in which they are analyzed (Validation & Compliance Institute 2024).
    - Identification of critical process parameters: Identifying parameters that significantly affect the quality of the final product is crucial. These may include factors or variables, such as time, temperature, pressure, and humidity (IIoT World 2017). Establishing critical limits for these parameters and ensuring that they are maintained within those limits during the process is vital for ensuring product quality and safety (Alhuraish et al. 2017).
    - Development of validation protocols: Validation protocols detail the specific methods and procedures to be followed during the validation process, which outlines the activities (Anju and Pandey 2017). This may include data collection, performance testing, statistical analysis, and other activities necessary to demonstrate the validity of the process (Sadiq et al. 2021).
    - Creating a detailed timeline: A detailed timeline is essential to ensure that all validation activities are performed promptly and efficiently (Validation & Compliance Institute 2024). This timeline will include deadlines for data collection, testing, and analysis, as well as a plan to review and approve the results or modifications if needed (Anju and Pandey 2017).
4. Perform process validation: In this stage, the validation plan is carried out step by step, and the results obtained are recorded in the formats created for this purpose (Calvo-Manzano et al. 2016).

5. Analyze results and make adjustments as necessary: After completing process validation, it is important to analyze the results obtained and assess whether the established acceptance criteria are met (Zhao et al. 2017). If the results do not comply, process adjustments may be necessary to improve performance and meet established standards in consensus with the area involved (Ika 2009).
6. Document and report results: It is critical that all findings, conclusions, and actions taken during process validation be accurately documented and retained for further study and analysis (Sandhya et al. 2015). This includes the procedures used, the results obtained, and the corrective or preventive actions implemented (Brennan 2014).

1.5 BARRIERS TO THE IMPLEMENTATION OF PROCESS VALIDATION

There is no doubt that process validation is crucial to ensure quality and consistency in product manufacturing; however, there are several barriers that can hinder its implementation or require a greater number of resources. The following is a list of common barriers to implementing process validation in manufacturing:

- Lack of understanding and awareness: One of the primary challenges in implementing process validation in manufacturing is the lack of understanding of its importance by top management and other departments involved (Zhao et al. 2017). Process validation is sometimes perceived as an additional expenditure of time and resources rather than an investment in product quality and customer satisfaction, which seeks to identify areas of opportunity for improvement (Sandhya et al. 2015).
- Resistance to change: Another common barrier is employee resistance to change, where they do not accept modifications and improvements to the methods and procedures they have already mastered (Sandhya et al. 2015). Implementing process validation often requires the adoption of new practices and procedures, which can generate resistance among staff who are comfortable with the way tasks have been performed, even if the quality is insufficient (Gonçalves and Gonçalves 2012).
- Lack of resources: Proper process validation implementation requires financial, human, and technological resources. Therefore, a lack of adequate resources, including trained personnel, technology,

and equipment, can hinder the effective implementation of process validation (Anju and Pandey 2017). Without adequate personnel and necessary tools, it is difficult to carry out a rigorous validation process efficiently (Sandhya et al. 2015). Managerial commitment to the production process is a solution to many of these problems.

- Standards and regulations: It is also important to note regulatory barriers and internal and external regulatory requirements (Rommerskirchen 2015). In some cases, regulations can be ambiguous or difficult to comply with, which makes the process validation process difficult and can lead to different interpretations among the different parties involved (Mathiyazhagan et al. 2022).
- Organizational culture: The lack of a quality-oriented culture throughout the organization can be a significant barrier, starting with a managerial focus (McKinsey & Company 2017). If the company's mindset does not emphasize the importance of quality and continuous improvement, process validation can be perceived as a hindrance rather than a tool to improve quality and efficiency. Hence, the importance of management promotes and raises awareness among its employees to ensure established benchmarks (Anju and Pandey 2017).
- Process complexity: Some organizations may have complex and highly interconnected processes, which makes it difficult to implement process validation as several areas are involved (Anju and Pandey 2017). Process validation requires a thorough understanding of existing processes and the ability to identify and address potential risks and deviations. Process complexity can make validation more labor-intensive and requires more effort to ensure that it is performed properly, integrating the components involved (Anta et al. 2022).
- Difficulties in data collection and management: Process validation requires collecting, analyzing, and managing accurate and reliable data (Rommerskirchen 2015). However, many organizations may face difficulties in obtaining and managing such data. This may be due to a lack of adequate systems for collecting and storing data, as well as problems with the quality and consistency of the available data (Townsend et al. 2018). Undoubtedly, data that do not represent the process can lead to incorrect decisions.
- Lack of equipment and technology: Process validation may require significant investment in equipment and technology acquisition and staff training (Anju and Pandey 2017). However, despite these difficulties and limitations, process validation is critical for ensuring the quality and safety of an organization's products and services.

- Lack of top management support: Process validation may require major organizational procedures and practice changes (Sandhya et al. 2015). This may include implementing new technologies, reorganizing workflows, and allocating additional resources; for these, top management authorization is required. Thus, a lack of top management support and commitment may hinder the successful implementation of process validation and make it difficult to overcome the barriers mentioned above (Wilson 2019).

In summary, implementing process validation in manufacturing can face multiple barriers, including a lack of understanding of its importance to resistance to change, a lack of resources, regulatory barriers, and organizational culture. These challenges must be addressed strategically to achieve an effective and beneficial implementation for both the company and its customers. Still, it must be initiated by the management, who is responsible for disseminating quality culture.

1.6 CRITICAL FACTORS IN PROCESS VALIDATION IMPLEMENTATION

Validation of the manufacturing process is a crucial step in ensuring the quality and consistency of manufactured products. To effectively carry out this process, it is important to consider the following critical success factors:

- Management leadership and commitment must establish a clear vision, provide the necessary resources, and foster a culture of continuous improvement (Alnadi and McLaughlin 2020). As Alnadi and McLaughlin (2020) point out, leadership is a key element in facilitating the successful implementation of strategies, such as Lean Six Sigma, which aim to improve quality, reduce costs, and increase the efficiency of the production process (Sreedharan et al. 2018).
- Personnel involvement and training: The personnel in charge of manufacturing and quality control must be trained and committed to the validation process to understand the process (Ngadiman et al. 2017). This involves providing training in validation concepts, techniques, and tools, as well as fostering a continuous improvement of the mindset. As they indicate, employee involvement and cultural change are critical factors for implementing a lean philosophy, where process validation is vital (Mathiyazhagan et al. 2022).

- Use of improvement methodologies and tools: The application of improvement methodologies and tools, such as Lean Six Sigma, Quality by Design, and DOE, is fundamental to carrying out robust and systematic manufacturing process validation (Medium 2021). These techniques allow the identification and control of key process parameters, reduce variability, and ensure the quality of the final product (Anju and Pandey 2017). As Mathiyazhagan et al. (2022) mentioned, implementing lean manufacturing principles has proven effective in improving the efficiency of production processes in the automotive industry and can be applied across various sectors.
- Systems and data integration: Integrating quality, production, and data management systems is crucial for gaining a holistic view of the manufacturing process and facilitating evidence-based decision-making during validation (Bousdekis et al. 2022). Digitization and the use of tools such as building information modeling can contribute to this goal (ModelArk 2018). However, it should be mentioned that associating a parameter with certain processes and activities is a complex process.

This indicates that the critical success factors for conducting manufacturing process validation include management leadership and commitment, staff involvement and training, improvement methodologies and tools, and integrating systems and data. By effectively addressing these elements, organizations can improve the quality, efficiency, and reliability of their manufacturing processes.

1.7 PROCESS VALIDATION IMPLEMENTATION BENEFITS

Performing manufacturing process validation brings several significant benefits to companies seeking to ensure the quality and efficiency of their operations, some of which are discussed as follows:

- Reduced risk helps companies identify and mitigate potential risks in their manufacturing processes, improve the final product's safety and reliability (Ostrove 2016), and seek to reduce or eliminate them.
- Increased productivity by allowing processes to be optimized appropriately and documented, where companies can reduce waste, improve cycle times, and increase profitability (Bhatt 1995).

- Regulatory compliance with regulatory bodies requires the validation of manufacturers' manufacturing processes to demonstrate compliance with applicable standards and regulations, especially those of an environmental nature (Zhao et al. 2017).
- Quality improvement because validation also helps companies improve the quality and consistency of their products. By implementing a robust process validation program, companies can increase customer satisfaction, reduce warranty and repair costs, and establish a reputation for quality (Souto et al. 2007).

In summary, process validation is fundamental for companies seeking to optimize their operations, ensure quality, and comply with regulatory requirements.

1.8 PROCESS VALIDATION-A BIBLIOMETRIC LITERATURE REVIEW

A bibliometric review was conducted to understand how scientific and academic research on process validation has evolved. This analysis allowed us to identify the main authors, institutions of higher education, and countries researching this topic.

1.8.1 The Database and Search Equations

The Scopus database was chosen to identify papers on process validation because it integrates many prestigious publishers that meet specific quality requirements. The initial search equation was TITLE-ABS-KEY ("process validation") to identify documents containing these two complete words in the article title, abstract, or keywords. This search topic contained 1674 documents; however, many validated processes were associated with other industrial sectors, not manufacturing.

The term "word manufacturing" was added to ensure that it referred to an industrial process that required the manufacturing of some products. Thus, the search equation was TITLE-ABS-KEY ("process validation") AND TITLE-ABS-KEY (manufacturing)). This search resulted in 498 documents. However, 20 documents from 2024 and one from 2025 were discarded to avoid statistical bias. Thus, only 477 documents were analyzed, including different languages, such as English, French, German, Chinese, Spanish, Portuguese, Turkish, and Swedish.

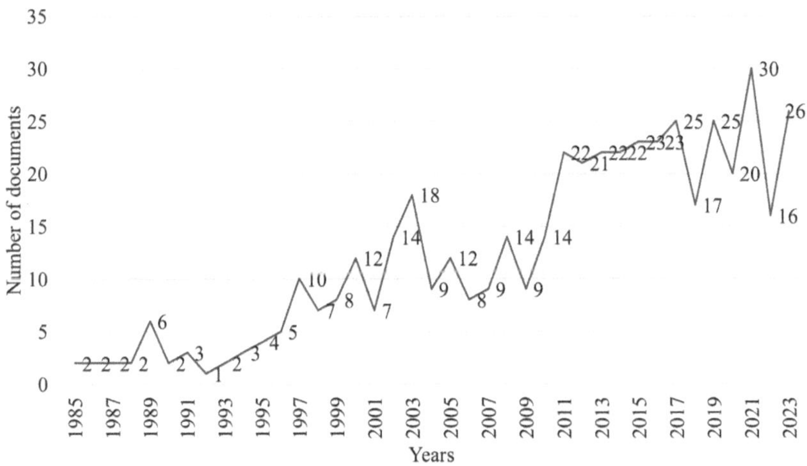

FIGURE 1.1 Trend in scientific publications on process validation in manufacturing.

1.8.2 Trend in Process Validation

Figure 1.1 indicates that process validation has appeared in academic reports since 1985, as seen in the conference report by Cattaneo (1985) and the scientific article by Jones and O'Connor (1985), published in a journal. Between 1985 and 1993, the number of published papers was much higher at approximately two per year. However, an increase was observed between 1993 and 2003, reaching 18. A decrease was observed from 2003 to 2010, after which it increased sharply. These documents are distributed as follows: journal (312), books (62), conference proceedings (62), trade journals (29), and book series (12).

In conclusion, the topic of process validation has been widely accepted in the scientific community in the manufacturing field, and the upward trend in the number of papers published proves this.

1.8.3 Areas Where Process Validation Is Published

A total of 23 areas in which process validation was published were identified, and it was observed that areas associated with pharmacology predominated. These areas are as follows: Pharmacology, Toxicology, and Pharmaceutics—244; Medicine—124; Biochemistry, Genetics, and Molecular Biology—112; Engineering—110; Chemistry—50; Immunology and Microbiology—44;

Materials Science—38; Chemical Engineering—37; Computer Science—35; Physics and Astronomy—23; Mathematics—16; Business, Management, and Accounting—15; Agricultural and Biological Sciences—7; Decision Sciences—4; Energy—4; Environmental Science—4; Health Professions—4; Earth and Planetary Sciences—3; Economics, Econometrics, and Finance—3; Nursing—3; Social Sciences—3; and Veterinary—3.

1.8.4 Top Journals Publishing on Process Validation

A total of 118 journals were identified as having published articles on the topic of process validation. A review of the names of these journals shows that many are in the areas of medicine, pharmaceuticals, and medical sciences. This may be due to the criticality of processes in this industrial sector, where manufacturing or manufacturing errors can have fatal consequences. The top 10 journals publishing on the topics of process validation are the following:

1. *STP Pharma Pratiques*, with 13 documents.
2. *Biopharm International*, with 12 documents.
3. *PDA Journal of Pharmaceutical Science and Technology*, with 11 documents.
4. *Drug Development and Industrial Pharmacy*, with ten documents.
5. *Pharmazeutische Industrie*, with ten documents.
6. *Pharmaceutical Engineering*, with eight documents.
7. *Cytotherapy*, with seven documents.
8. *Pharmaceutical Technology*, with seven documents.
9. *Biotechnology Progress*, with six documents.
10. *International Journal of Pharmtech Research*, with six documents.

1.8.5 Authors in the Area of Process Validation

At this point, the question is which authors have published in a journal. From the analysis of 477 papers, 159 authors were identified, with at least one paper on process validation. However, it is impossible to identify a group with consolidated lines of research because they are isolated. Table 1.2 indicates those authors who have at least two papers, and it is observed that the maximum number they have is only six.

TABLE 1.2 Authors on the topic of process validation

AUTHORS	PAPERS
Kirrstetter, R.; Rathore, A.S.	6
Jornitz, M.W.; Vishal Gupta, N.	5
Bracke, S.; Herwig, C.; Sofer, G.	4
Canfield, S.L.; Geigert, J.; Godavarti, R.; Krejci, S. Lau, A.S.L.; Lindboe, W.G.; Norling, L.A.; Seely, R.J. Shi, L.; Xu, Y.; Zahel, T.;	3
Antoni, G.; Backes, B.; Baseman, H.; Basso, A.; Blank, G.S.	2
Bussineau, C.; Calcott, P.H.; Campa, C.; Cappia, J.M.; Charlebois, T.S.; Chen, Q.; Chow, S.C.; Clenet, D.; Conrad, D.; DeSantis, P.; Dellepiane, N.; Desai, S.; Ding, W.; Doverfjord, J.G.; Drapeau, D.; Fabian, A.C.; Feuerhelm, D.; Garcia-Montoya, E.; Gastineau, T.; Germay, O.; Gervais, D.; Haury, J.; Helle, M.R.; Hentschel, N.; Huxsoll, J.F.; Inoue, M.; Juvin, P.; Kam, E.; Kao, Y.H.; Kelley, B.D.; Kind, S.; King, P.G.; Kohno, H.; Krause, S.O.; Krishnan, J.; Langley, D.; Levine, H.L.; Mallia-Milanes, A.; Mannermaa, J.P.; Martin, J.; Martin-Moe, S.; Max, S.I.; McGoldrick, M.; Meltzer, T.H.; Menezes, J.C.; Mlynarczyk, P.J.; Morefield, E.; Munyakazi, L.; Nash, R.A.; Niemann, M.; Notarnicola, S.; Novick, S.; O'Leary, R.	
O'Leary, R.M.; Ohtake, S.; Ostrove, S.; Pagliusi, S.; Patil, S.S.; Perez-Lozano, P.; Raghunandan, H.V.; Rao, G.; Reynier, J.P.; Rudge, S.; Rueckert, A.; Rustomjee, M.T.; Saltus, E.; Seely, J.E.; Seno, S.; Shibakov, A.; Sivaramakrishnan, V.H.; Skuballa, N.; Stark, R.; Stockdale, D.; Suñé-Negre, J.M.; Suñé-Pou, M.; Tetzlaff, R.F.; Thien, M.P.; Van Ooij, M.; Velikyan, I.; Werner, R.G.; Wilkinson, D.; Yamada, S.; Yusuf-Makagiansar, H.; Zabriskie, D.W.	

1.8.6 Institutions That Publish

It is crucial to understand the institutions to which these authors are affiliated. The main institutions with which these authors are affiliated, as well as the number of papers published on the subject of manufacturing process validation, are listed as follows:

1. Amgen Incorporated, with 16 documents.
2. Genentech Inc., with 16 documents.
3. Food and Drug Administration, with 12 documents.
4. Pfizer Inc., with 11 documents.
5. Biogen Inc., with 11 documents.
6. Merck & Co., Inc., with nine documents.

7. GlaxoSmithKline plc., with nine documents.
8. AstraZeneca, with eight documents.
9. Sanofi-Aventis Deutschland GmbH, with seven documents.
10. Novartis International AG, with five documents.
11. GlaxoSmithKline, USA, with five documents.
12. F. Hoffmann-La Roche AG, with four documents.
13. Johnson & Johnson, with four documents.
14. Boehringer Ingelheim International GmbH, with four documents.
15. Indian Institute of Technology Delhi, with four documents.

This shows that the authors came from pharmaceutical institutions associated with the FDA. This underscores the critical importance of process validation in the industrial sector, where an error can have fatal consequences. From this analysis, some institutions are prolific in the topic of process validation, such as Amgen Incorporated and Genentech, Inc., which have 16 documents each and very clear procedures and methodologies to validate their processes. Specifically, seven institutions had four documents, 13 had three documents, 100 had two documents, and 63 had only one.

1.8.7 Countries in Which It Is Published

Fifty countries were identified where process validation research was conducted, and Table 1.3 illustrates this relationship. Forty documents were not

TABLE 1.3 Countries publishing on the topic of process validation

COUNTRY	*DOCUMENTS*
United States	209
India	59
Germany	54
United Kingdom	34
Switzerland	23
France	20
China, Italy	12
Canada	10
Sweden	9
Japan	8
Spain	8

(*Continued*)

TABLE 1.3 (Continued)

COUNTRY	*DOCUMENTS*
Ireland, the Netherlands	7
Austria, Brazil	6
Portugal	5
Belgium, Denmark	4
Australia, Finland, Turkey	3
Colombia, Czech Republic, Hungary, Luxembourg, Malaysia, Mexico, Norway, Singapore, Slovenia, South Africa, South Korea, South Africa	2
Argentina, Croatia, Cuba, Eritrea, Iran, Israel, New Zealand, Pakistan, Peru, Philippines, Poland, Puerto Rico, Romania, Saudi Arabia, Sri Lanka, Taiwan, Ukraine, Vietnam	1
Undefined	40

associated with any of the countries. The United States leads the number of publications with 209 papers, followed by India with 59 and Germany with 54.

It was expected that countries with large manufacturing centers, such as China and Mexico, would have more documents. Still, the first country would be in position 7, and the second in position 28. This may be because these countries only manufacture products, but do not design them because, as manufacturing centers, companies have their headquarters in other countries. The same problem can be seen in countries such as Japan and South Korea, where the former has generated only eight documents and the latter only two, and these are not manufacturing countries but are large producers. This may be because when discussing the validation of their industrial processes, they consider it a trade secret and do not publish it.

1.9 CONCLUSION

The validation of manufacturing processes is a crucial aspect in the industry to guarantee the quality and efficiency of products, so it is a fundamental step to ensure that the manufactured products meet the required standards. Through this process, it is possible to identify deviations in production that could affect the quality of the final product, as well as to establish preventive and corrective measures to guarantee manufacturing consistency. In addition, process validation allows for the optimization of efficiency and cost reduction by identifying opportunities for improvement in the production flow. It is important to note that this activity should be a continuous process with periodic reviews to ensure that established quality standards are maintained and that best manufacturing practices are incorporated into the operation.

Several important conclusions can be drawn after the manufacturing process validation, as mentioned as follows.

- First, validating a manufacturing process demonstrates that the process can produce products that meet established quality standards. This provides confidence in a company's ability to meet customer expectations and regulations.
- Second, by performing validation, possible deviations in the manufacturing process were identified, allowing the implementation of preventive and corrective measures to ensure the consistency and quality of the final product. These measures are essential to minimize the risk of nonconformity and ensure that manufactured products comply with the specifications.
- Third, process validation identifies opportunities for improvement in the production flow, leading to greater efficiency and cost reduction. A company can achieve savings and improve its competitiveness by optimizing its manufacturing processes.
- Fourth, the validation of manufacturing processes contributes to the identification of risk areas in production, which allows for the implementation of strategies to mitigate possible process failures and guarantee the safety of manufactured products.
- In addition, the importance of training personnel involved in manufacturing processes has become evident, as well-trained operators are essential for ensuring quality and consistency in production.
- Similarly, process validation provides a framework for continuous improvement, encouraging innovation and the adoption of new technologies that can improve efficiency and product quality.
- Finally, it is worth noting that the bibliometric analysis reveals that this topic to be academic and scientific interest, as the number of documents has increased over the years. Countries such as the United States and India have generated more documents, many of which are in the pharmaceutical industry sector, where industrial processes are critical.

In summary, validating the manufacturing process is fundamental for ensuring product production consistency, quality, and efficiency. It is a continuous process that requires constant review and improvement to adapt to industry changes and to continue to meet quality standards.

REFERENCES

Adams, C. A. 1919. "Industrial Standardization." *The ANNALS of the American Academy of Political and Social Science* 82 (1): 289–99. https://doi.org/10.1177/000271621908200130/ASSET/000271621908200130.FP.PNG_V03.

Advance Electrical Design & Engineering Institute. 2024. "Process Design Training Course." https://www.advanceelectricaldesign.com/Process-Design-Training-Course.

Alhuraish, Ibrahim, Christian Robledo, and Abdessamad Kobi. 2017. "A Comparative Exploration of Lean Manufacturing and Six Sigma in Terms of Their Critical Success Factors." *Journal of Cleaner Production* 164 (October): 325–37. https://doi.org/10.1016/J.JCLEPRO.2017.06.146.

Alnadi, Mohammad, and Patrick McLaughlin. 2020. "Leadership That Facilitates the Successful Implementation of Lean Six Sigma." In *IMMS '20: Proceedings of the 3rd International Conference on Information Management and Management Science*, 59–66. Association for Computing Machinery. https://doi.org/10.1145/3416028.3416045.

Anderson, Neal G., David C. Burdick, and Maxwell M. Reeve. 2011. "Current Practices of Process Validation for Drug Substances and Intermediates." *Organic Process Research and Development* 15 (1): 162–72. https://doi.org/10.1021/OP1002825/ASSET/IMAGES/MEDIUM/OP-2010-002825_0002.GIF.

Anju, Goyal, and Priyambada Pandey. 2017. "Process Validation of Pharmaceutical Dosages Form: A Review." *Biomedical Journal of Scientific & Technical Research* 1 (5): 1–9. https://doi.org/10.26717/BJSTR.2017.01.000463.

Anta, Antonio Fernández, Chryssis Georgiou, Nicolas Nicolaou, and Antonio Russo. 2022. "Validated Objects: Specification, Implementation, and Applications." *ArXiv* 1 (May): 1–19. https://doi.org/10.48550/arXiv.2205.13448.

Avramova, Tanya, Dimka Vasileva, and Teodora Peneva. 2024. "An Overview of the Basic Concepts and Terms Related to Manufacturing Process Capability Evaluation." In *AIP Conference Proceedings*, edited by Rosen Hristov and Radostin Dimitrov. Vol. 3104. Varna, Bulgaria: American Institute of Physics Inc. https://doi.org/10.1063/5.0198835/3265813.

Bai, Zhong Hang, Peng Zhang, Fang Liu, and Run Hua Tan. 2007. "Product Development Performance Measures in Manufacturing Firm." In *2007 IEEE International Conference on Industrial Engineering and Engineering Management*, edited by Martin Helander, Min Xie, Roger Jiao, and Kay Chuan Tan, 926–29. Singapore: IEEE. https://doi.org/10.1109/IEEM.2007.4419327.

Balamurali, S. 2022. "Combined Attri-Vari Inspection Policy for Resubmitted Lots Based on the Process Capability Index." *Communications in Statistics—Simulation and Computation* 51 (9): 5406–25. https://doi.org/10.1080/03610918.2020.1771367.

Bhatt, Vinay. 1995. "How Validation Pays for Itself." *Drug Information Journal* 29 (2): 675–81. https://doi.org/10.1177/009286159502900246.

Bousdekis, Alexandros, Katerina Lepenioti, Dimitris Apostolou, and Gregoris Mentzas. 2022. "Data Analytics in Quality 4.0: Literature Review and Future Research Directions." *International Journal of Computer Integrated Manufacturing* 36 (5): 678–701. https://doi.org/10.1080/0951192X.2022.2128219.

Brennan, Zachary. 2014. "EMA Finalizes Guidance on Process Validation for Manufacturers." https://www.outsourcing-pharma.com/Article/2014/03/04/EMA-finalizes-guidance-on-process-validation-for-manufacturers.

Calvo-Manzano, Jose, Gonzalo Cuevas Agustin, Tomas San Feliu Gilabert, and Jose De J. Jimenez Puello. 2016. "DEPROVAL:Herramienta Para La Definición Del Proceso de Validación En Pequeñas Organizaciones de Software." In *Iberian Conference on Information Systems and Technologies, CISTI*, July, 1–6. Gran Canaria, Spain: IEEE Computer Society. https://doi.org/10.1109/CISTI.2016.7521500.

Cattaneo, Donald J. 1985. "Health Car Industry Validation in the 1990'S." In *39th Annual Quality Congress Transactions*, 325–33. Baltimore, MD: ASQC. https://www.scopus.com/record/display.uri?eid=2-s2.0-0021835205&origin=inward&txGid=c512d05d1c01d915ee68c5d06b3659d9.

CEA Instrumentos. 2016. "Instrumentos Para Control de Calidad y Sistemas de Inspecciones Para Una Amplia Gama de Industrias." https://www.ceainstrumentos.cl/.

Colledani, Marcello, Tullio Tolio, Anath Fischer, Benoit Iung, Gisela Lanza, Robert Schmitt, and József Váncza. 2014. "Design and Management of Manufacturing Systems for Production Quality." *CIRP Annals—Manufacturing Technology* 63 (2): 773–96. https://doi.org/10.1016/j.cirp.2014.05.002.

De Mast, Jeroen, and Ronald J.M.M. Does. 2006. "Industrial Statistics: A Discipline with Opportunities and Challenges." *Statistica Neerlandica* 60 (3): 270–82. https://doi.org/10.1111/J.1467-9574.2006.00329.X.

De Ron, A. J., and J. E. Rooda. 2006. "OEE and Equipment Effectiveness: An Evaluation." *International Journal of Production Research* 44 (23): 4987–5003. https://doi.org/10.1080/00207540600573402.

de Toledo, Jose Carlos, Fabiane Letícia Lizarelli, and Manoel Bispo Santana Junior. 2017. "Success Factors in the Implementation of Statistical Process Control: Action Research in a Chemical Plant." *Production* 27 (June): e20162208. https://doi.org/10.1590/0103-6513.220816.

Dong, Li Ning, Yong Jun Ruan, Zhan Wei Liu, and Bo Yin. 2012. "Research on the Methods of Precision Equipment Maintenance Based on CBM." In *Proceeding of 2012 International Conference on Information Management, Innovation Management and Industrial Engineering, ICIII 2012*, edited by Lin Yang, Edwin Cheung, and Yang Yongsheng. Vol. 2, 88–91. Sanya, China: IEEE Computer Society. https://doi.org/10.1109/ICIII.2012.6339784.

Dudek-Burlikowska, M. 2005. "Quality Estimation of Process with Usage Control Charts Type X-R and Quality Capability of Process Cp, Cpk." *Journal of Materials Processing Technology* 162–163 (Spec. Iss.): 736–43. https://doi.org/10.1016/J.JMATPROTEC.2005.02.210.

Encyclopedia.com. 2018. "Quality Control." https://www.encyclopedia.com/social-sciences-and-law/economics-business-and-labor/businesses-and-occupations/quality-control.

GMP7.COM. 2019. "Drug Product Manufacturing—GMP Validation Master Plan (VMP)." https://www.gmp7.com/validation-master-plan-drug-product-manufacturing-gmp-vmp-validation-master-plan.

Gonçalves, Joel Mana, and Rejane Pereira da Silva Gonçalves. 2012. "Overcoming Resistance to Changes in Information Technology Organizations." *Procedia Technology* 5 (January): 293–301. https://doi.org/10.1016/J.PROTCY.2012.09.032.

Grace, Ernest, Rosita Manawari Girsang, Sudung Simatupang, Vivi Candra, Novelyn Sidabutar, Sekolah Tinggi, Ilmu Ekonomi, and Sultan Agung. 2021. "Product Quality and Customer Satisfaction and Their Effect on Consumer Loyalty." *International Journal of Social Science* 1 (2): 69–78. https://doi.org/10.53625/IJSS.V1I2.138.

Grozdić, Vanja, and Jelena Demko-Rihter. 2023. "Economic Evaluation of Investment Projects: Determining the Key Factors for Final Investment Decision." *Journal of Local Self-Government* 21 (1): 45–70. https://doi.org/10.4335/21.1.45-70(2023).

Guo, Hongfei, Ru Zhang, Min Zou, Junda Yang, Jingge Zhang, Zhihui He, Ting Qu, Congdong Li, and Guoquan Huang. 2018. "Review on Quality Management Model Based on Process Approach." *Business and Management Horizons* 6 (2): 26–33. https://doi.org/10.5296/BMH.V6I2.13955.

Gutiérrez Pulido, Humberto, and Román de la Vara Salazar. 2013. *Control Estadístico de La Calidad y Seis Sigma*. 3rd ed. Mexico City: McGraw-Hill.

HEFLO. 2024. "How to Implement Business Process Improvement Analysis?" *Best Practices to Analyze and Improve Your Business Process*. https://www.heflo.com/blog/process-optimization/business-process-improvement-analysis/.

IIoT World. 2017. "Process Monitoring in the Automotive Industry." https://www.iiot-world.com/industrial-iot/connected-industry/process-monitoring-in-the-automotive-industry/.

Ika, Lavagnon A. 2009. "Project Success as a Topic in Project Management Journals." *Project Management Journal* 40 (4): 6–19. https://doi.org/10.1002/PMJ.20137.

Jatto, Elsie, and Augustine O. Okhamafe. 2002. "An Overview of Pharmaceutical Validation and Process Controls in Drug Development." *Tropical Journal of Pharmaceutical Research* 1 (2): 115–22. https://tspace.library.utoronto.ca/bitstream/1807/1485/1/pr02016.pdf.

Jones, A. J., and J. V. O'Connor. 1985. "Control of Recombinant DNA Produced Pharmaceuticals by a Combination of Process Validation and Final Product Specifications." *Developments in Biological Standardization* 59 (January): 175–80. https://europepmc.org/article/med/4007276.

Kenyon, George N., and Kabir C. Sen. 2015. *The Perception of Quality: Mapping Product and Service Quality to Consumer Perceptions. The Perception of Quality*. 1st ed. London: Springer. https://doi.org/10.1007/978-1-4471-6627-6.

Kneat. 2024. "Digital Process Validation for Faster Speed to Market." *Why Kneat Use Proven Technology and Validation Industry Expertise to Support Your Process Validation Program*. https://kneat.com/application/process-validation/.

Korzenowski, A. L., G. Vidor, G. L.R. Vaccaro, and C. S. Ten Caten. 2015. "Control Charts for Flexible and Multi-Variety Production Systems." *Computers & Industrial Engineering* 88 (October): 284–92. https://doi.org/10.1016/J.CIE.2015.07.016.

Lam, Herman. 2004. "Procurement, Qualification, and Calibration of Laboratory Instruments: An Overview." In *Analytical Method Validation and Instrument Performance Verification*, January, 139–51. https://doi.org/10.1002/0471463728.CH9.

Leclair, J. A. 2013. "Training in the Nuclear Industry." In *Managing Nuclear Projects*, edited by Jas Devgun, 41–67. Woodhead Publishing. https://doi.org/10.1533/9780857097262.1.41.

Lee, Jong-Min, Chang Kyoo Yoo, and In-Beum Lee. 2004. "Statistical Process Monitoring with Independent Component Analysis." *Journal of Process Control* 14: 467–85. https://doi.org/10.1016/j.jprocont.2003.09.004.

Lina, Roidah. 2022. "Improving Product Quality and Satisfaction as Fundamental Strategies in Strengthening Customer Loyalty." *AKADEMIK: Jurnal Mahasiswa Ekonomi & Bisnis* 2 (1): 19–26. https://doi.org/10.37481/JMEB.V2I1.245.

Maishi MFG. 2022. "Quality Control System." https://www.maishimfg.com/running/quality-control-system.html.

Mathiyazhagan, K., A. Gnanavelbabu, Naveen Kumar.N, and Vernika Agarwal. 2022. "A Framework for Implementing Sustainable Lean Manufacturing in the Electrical and Electronics Component Manufacturing Industry: An Emerging Economies Country Perspective." *Journal of Cleaner Production* 334 (February): 130169. https://doi.org/10.1016/J.JCLEPRO.2021.130169.

McKinsey & Company. 2017. "Manufacturing Quality Today: Higher Quality Output, Lower Cost of Quality." https://www.mckinsey.com/capabilities/operations/our-insights/manufacturing-quality-today-higher-quality-output-lower-cost-of-quality.

Medium. 2021. "Manufacturing Handbook of Best Practices: An Innovation, Productivity, and Quality Focus (St. Lucie Press/Apics Series on Resource Management)." https://medium.com/@myla428/pdf-download-manufacturing-handbook-of-best-practices-an-innovation-productivity-and-quality-99facd-4c55b6.

ModelArk. 2018. "Proyecta Inteligente, Construye Mejor." https://www.modelarkbim.com/.

Modgil, Sachin, and Sanjay Sharma. 2016. "Total Productive Maintenance, Total Quality Management and Operational Performance an Empirical Study of Indian Pharmaceutical Industry." *Journal of Quality in Maintenance Engineering* 22 (4): 353–77. https://doi.org/10.1108/JQME-10-2015-0048/FULL/XML.

Ngadiman, Yunos, Burairah Hussin, Abdul Talib Bon, and Nor Aziati Abdul Hamid. 2017. "Factors That Influenced the Quality Inspection on the Production Line in Manufacturing Industry." In *MATEC Web of Conferences: 2016 the 3rd International Conference on Mechatronics and Mechanical Engineering (ICMME 2016)*, edited by H.L. Yuan, R.K. Agarwal, P. Tandon, and E.X. Wang. Vol. 95, 1–4. Shanghai, China: EDP Sciences. https://doi.org/10.1051/MATECCONF/20179510007.

Ostrove, Steven A. 2016. *The Validation Life Cycle and Change Control. How to Validate a Pharmaceutical Process*. Academic Press. https://doi.org/10.1016/B978-0-12-804148-2.00003-2.

Partala, Timo, and Timo Saari. 2015. "Understanding the Most Influential User Experiences in Successful and Unsuccessful Technology Adoptions." *Computers in Human Behavior* 53 (December): 381–95. https://doi.org/10.1016/J.CHB.2015.07.012.

Pazhayattil, Ajay Babu, Naheed Sayeed-Desta, Emilija Fredro-Kumbaradzi, and Jordan Collins. 2018. "Life Cycle Approach to Process Validation." In *Solid Oral Dose Process Validation*. 1st ed., 1–7. Cham, Switzerland: Springer. https://doi.org/10.1007/978-3-030-02472-7_1.

Pharma Times Latest Drug Updates. 2023. "Guide for Pharmaceutical Manufacturing Quality Assurance New Drug Research." https://www.pharmacistspharma journal.org/.

PIA Automation. 2024. "Machine and System Qualification." https://www.piagroup.com/en/portfolio/machine-and-system-qualification/.

Prism Group of Company. 2024. "Prism Calibration Centre—Best Calibration Center in Gujarat." *Prism Calibration Centre*. https://prismcalibration.com/.

Psomas, Evangelos L., and Carmen Jaca. 2016. "The Impact of Total Quality Management on Service Company Performance: Evidence from Spain." *International Journal of Quality and Reliability Management* 33 (3): 380–98. https://doi.org/10.1108/IJQRM-07-2014-0090/FULL/XML.

Purba, H. H., A. Nindiani, A. Trimarjoko, C. Jaqin, S. Hasibuan, and S. Tampubolon. 2021. "Increasing Sigma Levels in Productivity Improvement and Industrial Sustainability with Six Sigma Methods in Manufacturing Industry: A Systematic Literature Review." *Advances in Production Engineering And Management* 16 (3): 307–25. https://doi.org/10.14743/APEM2021.3.402.

QMPG. 2017. "ISO/TS 16949 Automobile Certification Services in Kerala." *ISO/TS 16949(Automobile)*. https://qmpg.net/quality-management-systemqms/isots-16949automobile/.

Quality International. 2024. "Quality International." *IS Certification, ISO Certification, Calibrations, Third Party Inspections*. https://qualityinternational.org/.

Realyvásquez-Vargas, Arturo, Jorge Luis García-Alcaraz, Guadalupe Hernández-Escobedo, Karina Cecilia Arredondo-Soto, Joel Eduardo García-Ortíz, Julio Blanco-Fernández, and Emilio Jiménez-Macías. 2021. "The DMAIC Methodology as a Tool for Process Improvement: The Case of a Mexican Manufacturing Company." In *Trends in Industrial Engineering Applications to Manufacturing Process*, edited by J.L. García-Alcaraz, A. Realyvásquez-Vargas, and E. Z-Flores. 1st ed., 335–64. Cham, Switzerland: Springer Nature. https://doi.org/10.1007/978-3-030-71579-3_15.

Robinson, Sionade, and Lyn Etherington. 2006. *Customer Loyalty: A Guide for Time Travelers*. London: Palgrave Macmillan. https://doi.org/10.1057/9780230513037.

Rommerskirchen, Michael. 2015. "Continued Process Verification—a Challenge for the Pharmaceutical Industry?" *European Pharmaceutical Review* 20 (1): 48–51. https://www.europeanpharmaceuticalreview.com/article/29903/continued-process-verification-a-challenge-for-the-pharmaceutical-industry/.

Ryabchik, Tatiana A., Elvira E. Smirnova, Marina I. Lukashova, and Hasan Haydar. 2019. "Manufacturing Processes Quality Control as a Main Factor of Performance Enhancement in Industrial Management." In *2019 IEEE Conference of Russian Young Researchers in Electrical and Electronic Engineering (ElConRus 2019)*,

1463–66. St. Petersburg and Moscow: Institute of Electrical and Electronics Engineers Inc. https://doi.org/10.1109/EICONRUS.2019.8657176.

Sadiq, Saba, Muhammad Saad Amjad, Muhammad Zeeshan Rafique, Shafqat Hussain, Uzma Yasmeen, and Mohammad Aamir Khan. 2021. "An Integrated Framework for Lean Manufacturing in Relation with Blue Ocean Manufacturing—A Case Study." *Journal of Cleaner Production* 279 (January): 123790. https://doi.org/10.1016/J.JCLEPRO.2020.123790.

Sandhya, Ch., Brahmaiah Bonthagarala, Pusuluri Dharani Lakshmi Sai, and Konkipudi Venkata Sivaiah. 2015. "Process Validation: An Essential Process in Pharmaceutical Industry." *International Journal of Advances in Scientific Research* 1 (4): 179–82. https://doi.org/10.7439/ijasr.v1i4.1781.

Saravacos, George, and Athanasios E. Kostaropoulos. 2016. "Design and Selection of Food Processing Equipment." In *Handbook of Food Processing Equipment*, 51–85. Cham: Springer. https://doi.org/10.1007/978-3-319-25020-5_2/TABLES/4.

Sastri, Vinny R. 2014. "Process Validation for Medical Device Manufacturers and Their Suppliers." In *Plastics in Medical Devices*, edited by Vinny R. Sastri. 2nd ed., 279–94. MA: William Andrew Publishing. https://doi.org/10.1016/B978-1-4557-3201-2.00011-2.

Software Testing Company. 2022. "Qa Software Testing." *Providing QA Testing Services for Any Software Product*. https://qasoftwaretesting.com/.

Souto, E. B., T. Vasconcelos, D. C. Ferreira, and B. Saramento. 2007. "Pharmaceutical Manufacturing Validation Principles." In *Pharmaceutical Manufacturing Handbook: Regulations and Quality*, edited by Shayne Cox Gad, 811–38. NJ: John Wiley & Sons, Ltd. https://doi.org/10.1002/9780470259832.CH25.

Sreedharan, Raja V., Vijaya M. Sunder, and R. Raju. 2018. "Critical Success Factors of TQM, Six Sigma, Lean and Lean Six Sigma: A Literature Review and Key Findings." *Benchmarking: An International Journal* 25 (9): 3479–3504. https://doi.org/10.1108/BIJ-08-2017-0223/FULL/XML.

Teplická, Katarína, Samer Khouri, Tawfik Mudarri, and Magdaléna Freňáková. 2023. "Improving the Quality of Automotive Components Through the Effective Management of Complaints in Industry 4.0." *Applied Sciences* 13 (14): 8402. https://doi.org/10.3390/APP13148402.

Teuchert, V. 2015. "Validation of Process Control Systems in the Pharmaceutical Industry." In *European Control Conference, ECC 1999—Conference Proceedings*, 4818–23. Karlsruhe, Germany: Institute of Electrical and Electronics Engineers Inc. https://doi.org/10.23919/ECC.1999.7100099.

Tonini, Antonio Carlos, Fernando José Barbin Laurindo, and Mauro Mesquita De Spinola. 2007. "An Application of Six Sigma with Lean Production Practices for Identifying Common Causes of Software Process Variability." In *PICMET '07–2007 Portland International Conference on Management of Engineering and Technology*, edited by Dundar F. Kocaoglu, Timothy R. Anderson, and Tugrul U. Daim, 2482–90. Portland, OR: IEEE. https://doi.org/10.1109/PICMET.2007.4349584.

Townsend, Malcolm, Thanh Le Quoc, Gaurav Kapoor, Hao Hu, Wei Zhou, and Selwyn Piramuthu. 2018. "Real-Time Business Data Acquisition: How Frequent Is Frequent Enough?" *Information & Management* 55 (4): 422–29. https://doi.org/10.1016/J.IM.2017.10.002.

Tripathi, Varun, Somnath Chattopadhyaya, Alok K. Mukhopadhyay, Shubham Sharma, Jujhar Singh, Danil Yurievich Pimenov, and Khaled Giasin. 2021. "An Innovative Agile Model of Smart Lean–Green Approach for Sustainability Enhancement in Industry 4.0." *Journal of Open Innovation: Technology, Market, and Complexity* 7 (4): 215. https://doi.org/10.3390/JOITMC7040215.

Validation & Compliance Institute. 2024. "What Is Validation?" https://www.vcillc.com/about-vci/what-validation.

Walsh, Andrew, Tomohiro Doi, Kosuke Takenaka, Takashi Kaminagayoshi, Shunsuke Omori, Sadamitsu Makoto, Yoshiaki Miko, and Osamu Shirokizawa. 2018. "The Statistical Evaluation of Cleaning Processes Using Process Capability and Its Application to New Product Introduction." *Pharmaceutical Technology* 42 (5): 48–51. https://www.pharmtech.com/view/statistical-evaluation-cleaning-processes-using-process-capability-and-its-application-new-product-i.

Wang, Victoria, Kanak Agarwal, Sani R. Nassif, Kevin J. Nowka, and Dejan Marković. 2009. "A Simplified Design Model for Random Process Variability." *IEEE Transactions on Semiconductor Manufacturing* 22 (1): 12–21. https://doi.org/10.1109/TSM.2008.2011630.

Wang, Yuquan, Wei Wu, and Fusheng Liu. 2012. "The Quality Management Issues Emphasize in the Equipment Operational Phase." In *Proceedings of 2012 International Conference on Quality, Reliability, Risk, Maintenance, and Safety Engineering, ICQR2MSE 2012*, edited by Hong-Zhong Huang, Ming J. Zuo, and Yu Liu, 1425–28. Chengdu, China: IEEE. https://doi.org/10.1109/ICQR2MSE.2012.6246490.

Weiss, Brian A., Gregory Vogl, Moneer Helu, Guixiu Qiao, Joan Pellegrino, Mauricio Justiniano, and Anand Raghunathan. 2015. "Measurement Science for Prognostics and Health Management for Smart Manufacturing Systems: Key Findings from a Roadmapping Workshop." In *Annual Conference of the PHM Society*. Vol. 7, 323–33. Prognostics and Health Management Society. https://doi.org/10.36001/PHMCONF.2015.V7I1.2712.

Wilson, Charles. 2019. "How to Identify Roles and Responsibilities in a QMS Based on ISO 13485?" https://www.linkedin.com/pulse/how-identify-roles-responsibilities-qms-based-iso-13485-wilson.

Yao, Jinyong, Bobo Xiong, Yun Zhou, and Yihai He. 2017. "Research on Relationship between Manufacturing Process Quality Variation and Product Reliability." In *Proceedings of 2016 Prognostics and System Health Management Conference, PHM-Chengdu 2016*, edited by Ming J Zuo, Liudong Xing, Zhaojun Li, Zhigang Tian, and Qiang Miao, 1–6. Chengdu, China: Institute of Electrical and Electronics Engineers Inc. https://doi.org/10.1109/PHM.2016.7819909.

Zhao, Yincheng, Kuangjie Sheng, Zheng Wang, Xilin Zhang, H. Hengyiyang, and Rui Miao. 2017. "Process Validation and Revalidation in Medical Device Production." *Procedia Engineering* 174 (January): 686–92. https://doi.org/10.1016/J.PROENG.2017.01.207.

Auxiliary Methods and Tools in Process Validation 2

2.1 INTRODUCTION

Process validation is a crucial aspect of modern industry, especially in the context of lean manufacturing. In this regard, robust auxiliary methods and tools can make a difference in operational efficiency and final product quality. In the following sections, we explore some of the most effective methods and tools for process validation, with the objective of optimizing the results and maximizing the value delivered to the customer.

One of these is the 8Ds methodology, which is a powerful tool for solving problems, improving processes, and ensuring quality in various industries (Pazhayattil et al. 2018). Originating in the automotive sector, this methodology has spread to other areas due to its effectiveness in problem-solving (Omnex 2024). On the other hand, in this same context, the operation flow diagram becomes an invaluable tool for analyzing, understanding, and optimizing production processes (Shazia Sadiq et al. 2004). Similarly, the use of the process flow diagram as an auxiliary tool in process validation can provide a clear and detailed view of the activities involved, identify possible bottlenecks and critical points, and optimize operational efficiency (PMOinformatica.com 2015; Tripathi et al. 2021).

The Pareto diagram, a strategic tool, is invaluable for analyzing and visualizing the most significant problems in the validation process (Wilkinson 2006). By prioritizing these problems, companies can focus their resources on areas that significantly impact process improvement. Another crucial tool

DOI: 10.1201/9781003650966-2

in process validation is the Ishikawa diagram, also known as the fishbone diagram or cause-and-effect diagram (Bose 2012). This tool, developed by Professor Kaoru Ishikawa, enables the identification and analysis of the possible causes of a problem or undesired outcomes in a manufacturing process (Six Sigma 2017). By using the Ishikawa diagram in process validation, it is possible to visualize in a clear and structured way the different causes that could be affecting the quality or efficiency of the process (Luca and Luca 2019). This facilitates the identification of areas for improvement and the implementation of specific corrective actions to address each of the identified causes.

Finally, DOE has become a crucial tool in process validation, as it allows the identification and optimization of the factors that impact the quality and efficiency of the production process (Tripathi et al. 2021). By using DOE effectively, companies can reduce their development time, minimize costs, and improve quality.

This chapter explores how the 8Ds methodology can serve as an invaluable auxiliary tool for process validation. In addition, we will examine how the operation flow diagram acts as an auxiliary tool in process validation, contributing to the successful implementation of lean manufacturing strategies. We also explore the importance and impact of the process flow diagram in process validation within the manufacturing industry. We also explored how the Pareto diagram can be successfully applied in the process validation process, highlighting its importance and benefits in problem identification and resolution. We also explore how the Ishikawa diagram is used as an auxiliary tool in process validation in lean manufacturing and how its application can contribute to optimizing production processes. Finally, we explore how DOE can be applied in the context of process validation in lean manufacturing and how it can contribute to continuous improvement in the industry.

2.2 AUXILIARY TOOLS

2.2.1 8D Methodology

The 8D methodology is a structured and systematic approach in solving problems, identifying root causes, and taking corrective and preventive actions

(Realyvásquez-Vargas et al. 2020). This methodology consists of eight steps, which are summarized as follows:

1. Create teamwork: Create a multidisciplinary team that includes people with different skills and knowledge relevant to solving a problem (Carmignani 2009).
2. Identify the problem: Clearly define the problem or nonconformity to be solved (Øvretveit 1995).
3. Develop an interim plan: The team should establish a preliminary plan to address the problem, identify immediate actions to mitigate any negative impacts, and establish a timeline for the remaining actions (Tripathi et al. 2021). This plan should include a clear assignment of responsibilities to the team members (Ando 2024).
4. Perform root cause analysis: The team should use tools such as the Ishikawa diagram, the 5 Whys, and the Pareto diagram to identify possible root causes of the problem (Haughey 2023).
5. Analysis and selection of corrective actions: After identifying the root causes of the problem, the team must evaluate and select the most appropriate corrective actions to address each one (Six Sigma 2024). This may include process changes, additional training, product design improvements, or other necessary actions (QIT Consulting 2024).
6. Verify the effectiveness of corrective actions: Once implemented, the team must verify whether they have had the desired effect in resolving the problem (Rodriguez 2016).
7. Implementation of preventive actions: To avoid the recurrence of the problem, the team must identify and implement preventive actions (Dixon 2022). These actions may be related to updating procedures, quality control improvements, ongoing training, or other measures that will help prevent similar problems in the future (ASTM International 2019).
8. Implement closure actions and celebration: Once the problem has been effectively resolved, the team should document all actions taken and the results obtained, and hold a celebration to recognize the efforts and contributions of all team members (Leapsome 2024). This information is relevant for implementing the 8D approach for resolving problems or defects in a process or product (Mathiyazhagan et al. 2022).

In the context of process validation, the 8Ds methodology has proven to be an effective tool for identifying and resolving quality issues in a systematic manner (Anju and Pandey 2017), as well as problems in the validation process (Realyvásquez-Vargas et al. 2020). This approach allowed us to comprehensively address the various stages of process validation, thereby contributing to continuous improvement and compliance with quality standards (Pazhayattil et al. 2018). The importance of the 8Ds in process validation lies in their ability to effectively manage nonconformance situations (Muncut et al. 2019). In addition, applying 8Ds favors the early detection of deviations in validation processes, which allows the prompt implementation of corrective actions (Muncut et al. 2019; Realyvásquez-Vargas et al. 2020).

However, the literature highlights the relevance of the 8Ds methodology in improving the efficiency and effectiveness of validation processes by facilitating the identification of root causes and the definition of preventive actions. These findings support the importance of the 8Ds methodology as a critical tool in process validation, ensuring quality and compliance with regulatory requirements (Muncut et al. 2019). The 8Ds methodology has been the subject of numerous studies highlighting its usefulness in process validation (F. Zhou et al. 2016). The 8Ds provide a comprehensive framework for process validation problem-solving, leaving no stones unturned and contributing to cost reduction and increased operational efficiency. In addition, applying 8Ds promotes an organizational culture oriented toward continuous improvement, which is essential in highly regulated environments (Realyvásquez-Vargas et al. 2020).

Another relevant aspect is the ability of the 8Ds methodology to foster interdepartmental collaboration in the context of process validation (Muncut et al. 2019). This collaboration is not only beneficial but also critical for addressing complex challenges involving multiple areas of the organization, and the 8Ds provide a structured framework for this purpose (Calvo-Manzano et al. 2016). For their part, Sánchez Esquerra and Lozano Taylor (2001) pointed out that the 8Ds can be integrated with data analysis tools and process control techniques, highlighting their synergy with approaches based on statistical improvement (Realyvásquez-Vargas et al. 2020). This integration allows for a more robust approach to process validation, ensuring greater accuracy in problem identification and resolution (Realyvásquez-Vargas et al. 2020). Table 2.1 shows a brief literature review of the application of the 8Ds methodology in process validation.

To sum up, the 8Ds approach is a powerful instrument for process validation that helps find, analyze, and fix quality issues in a methodical way. In addition, its structured approach and eight key steps enable a comprehensive address of the stages of process validation, contributing to continuous improvement and compliance with quality standards. The 8Ds methodology is widely recognized as a practical approach to quality problem management and process validation improvement in the specialized literature.

TABLE 2.1 Application of 8Ds methodology for process validation

AUTHOR	*IMPLEMENTATION OF 8DS*	*RESULTS*
(Mitreva et al. 2015)	The research utilizes the 8Ds methodology to address the issue of an LED that is not functioning properly on the circuit board.	Employees' responsibility, participation and commitment toward quality and business processes were improved, and defects decreased.
(Bremmer 2015)	The study examines Scania's international supply chain and assesses strategies for ensuring product quality through the implementation of 8D methodologies and other approaches.	The production process is working but requires some improvements, mainly due to the expected growth of the North northbound flow.
(Zasadzień 2016)	The research seeks to solve problems identified in the railway carriage renovation process by implementing the 8Ds method.	The author utilized the 8Ds method to pinpoint the root causes of issues in the railway repair procedure and to devise corrective measures, leading to a significant enhancement in the efficiency of the process under scrutiny.
(Mitreva et al. 2016)	This study examines how well a car company's quality control system works. The authors looked at how the company manages its business processes, including identifying, documenting, and controlling them. They checked if the system's efficiency documentation was done correctly.	Quality and productivity at the lowest operating costs were defined.
(Titu 2016)	This study relies on the 8D method to solve a complaint about a defective part. It is conducted at SC COMPA S.A., a company based in Sibiu, Romania.	Sixty days after making changes to fix the issue, no more items were found with the same problem. Therefore, the client chose to cancel the complaints.

(*Continued*)

TABLE 2.1 (Continued)

AUTHOR	*IMPLEMENTATION OF 8DS*	*RESULTS*
(F. Zhou et al. 2016)	The study creates a way to improve quality for car companies using quality management methods. The 8D method and Six Sigma pilot programs were put into place.	The findings demonstrated that the proposed approach is impactful within the Chinese and South African automobile sectors evaluated.
(Nicolae et al. 2017)	This study presents a solution to reduce the response time for the 8Ds approach by: (a) warning workstations and warehouses about the appearance of a customer complaint, and (b) using a software program to manage some documents needed for the 8D analysis.	Because the problem is resolved when a customer complaint is received, one of the most important outcomes is a reduction in the time spent on communication between the quality team and the individuals involved in the production process. As a result of fixing the problem in response to consumer complaints, the time spent communicating between the quality team and those involved in production has markedly decreased. Among the most significant results is this: A novel consumer interface informs clients about the problem-solving measures being implemented. The platform adheres more closely to the conventional protocols for reporting 8D findings.
(Kumar and Adaveesh 2017)	The spring and stamping company conducted the six-month study. The research revealed a high rejection rate of valve springs (17.07%) as a result of defects. Consequently, the rate was decreased by 4.91% using the 8D procedure.	The product rejection rate decreased significantly in six months, from 17.07% in January 2014 to 4.91% in July 2014.

(*Continued*)

TABLE 2.1 (Continued)

AUTHOR	*IMPLEMENTATION OF 8DS*	*RESULTS*
(Roque and Berenice 2017)	The 8D method is employed in this study to develop and execute novel manufacturing processes for dental devices at Briggith using current technology. The objective is to guarantee the organization's continued existence in the present marketplace.	Standardizing raw materials and variables intervening in the process was possible. The advancement of modern, lightweight, and innovative architectural designs. The creation of a thorough electronic control system for the variables identified in the production process. The undertaking established compliance with quality standards.
(Škůrková 2017)	The research concentrates on reducing waste costs within an industrial organization. The author implemented the 8Ds method.	Identified were the causes of scrap costs, and corrective actions were implemented to reduce these costs.
(Wichawong and Chongstitvatana 2017)	In order to resolve the issue of abrupt stoppages in a continuous vacuum batch cooker utilized by a Brazilian sugar and alcohol company, this work employs the 8Ds method.	A pragmatic approach, integrated with high-quality tools, was employed to identify, resolve, and prevent the recurrence of the issue. The implementation of the 8Ds technique enhanced the company's performance and facilitated ongoing improvements in its production processes. Nonetheless, the strategy might also be utilized in other processes to enhance the company's competitiveness in quality and safety.

(Continued)

TABLE 2.1 (Continued)

AUTHOR	*IMPLEMENTATION OF 8DS*	*RESULTS*
(Vargas 2017)	This study applies the 8Ds method to address the issue of abrupt interruptions in a continuous vacuum batch cooker utilized by a Brazilian sugar and alcohol company.	A practical method combined with quality tools for detecting and solving the problem and eliminating its recurrence was implemented. The implementation of the 8Ds technique enhanced the company's performance and facilitated ongoing improvements in its production processes. Nonetheless, the strategy might also be utilized in other processes to enhance the company's competitiveness in quality and safety.
(Zasadzień 2017)	The research employs a quality engineering approach to enhance the maintenance procedures of a production facility in Silesia. The research employs the 8Ds technique to mitigate machine downtimes resulting from bottlenecks.	Machine downtimes caused by bottlenecks were significantly reduced.
(Pacheco-Pacheco 2018)	The study aims to enhance delivery times for modification clothes (Coarse High and Shirt High) products in a tailor shop through the application of the 8Ds methodology.	Production times diminished by 2.46% for Alto de Basta products (from 13.30 minutes to 12.98 minutes) and by 21.16% for Alto de Camisa products (from 8.49 minutes to 6.69 minutes). In both items, delivery wait times were reduced by 33.33%, decreasing from three days to two days.

Source: **Adapted from (Realyvásquez-Vargas et al. 2020)**

2.2.2 Operations Flowchart

An operation flowchart is a graphical representation that outlines the steps involved in a particular process or operation within an organization. It serves as a visual tool that delineates the flow of activities, decisions, and interactions among various system components, facilitating a clearer understanding of complex processes. In industrial applications, operation flowcharts are utilized to enhance process efficiency, ensure quality control, and streamline operations by providing a structured overview of workflows; however, they are also instrumental in various other ways. They help identify inefficiencies and bottlenecks within processes, enabling organizations to optimize their operations more effectively. For instance, the use of flowcharts in the manufacturing sector allows for the visualization of production processes, which aids in identifying redundant steps and the potential for waste reduction (Vidal et al. 2012). Furthermore, flowcharts can be integrated with Lean manufacturing and Six Sigma methodologies to support continuous improvement initiatives, thereby enhancing overall productivity (Cuenca et al. 2008).

Moreover, operation flowcharts are valuable in training and onboarding new employees. By providing a clear and concise depiction of processes, these charts serve as effective instructional tools that guide new hires through operational protocols (Martínez-Martínez 2023). In addition, flowcharts can be utilized in risk assessment and management, as they allow for the mapping of potential failure points and the development of contingency plans, particularly in high-stakes industries such as oil and gas (Rodríguez and Rodríguez 2004).

The versatility of operation flowcharts extends to various sectors, including healthcare, where they are employed to streamline patient care processes and improve service delivery (Martínez-Martínez 2023). In logistics and supply chain management, flowcharts facilitate the analysis of material flows, helping organizations to optimize their layouts and reduce transportation costs. Table 2.2 lists the symbols used in the flow diagram, whereas Figure 2.1 shows an example of a flow diagram representing the purchasing process for a shirt.

TABLE 2.2 Symbols used in a flowchart (LucidChart 2023)

OPERATION	*SYMBOL*	*DESCRIPTION*
Start/End	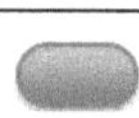	Indicates the beginning and end of the diagram.
Flow arrow		Marks the direction of data and process flow in the diagram.

(Continued)

TABLE 2.2 (Continued)

OPERATION	*SYMBOL*	*DESCRIPTION*
Process		This is where the operations or processes that will be performed with the input and output data will be obtained are expressed.
Data input/output		It is used to request data input or to display the outputs obtained.
Off-page connector		Point out that the continuation of the diagram flow continues on a different sheet.
In-page connector		Point out that the continuation of the diagram flow follows elsewhere on the same sheet.
Decision-making		It evaluates a condition and chooses one of two possible paths if it turns out to be true or false.

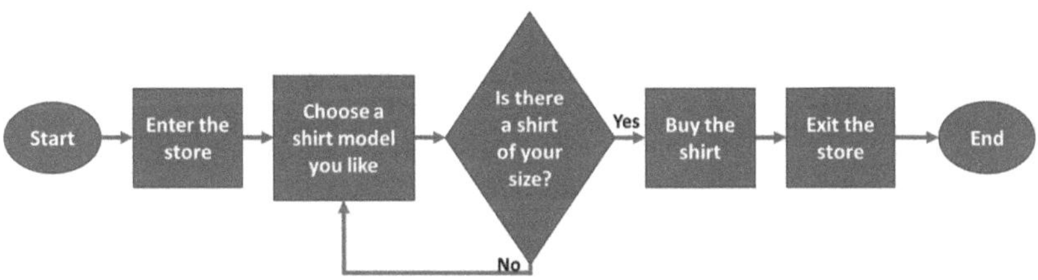

FIGURE 2.1 Example of a flow diagram

2.2.3 Process Flow Diagram

An operation flowchart is a graphical representation of the sequence of steps followed to carry out a specific process within an industrial operation (Y. Li et al. 2021). It is a visual tool that shows, in a clear and concise manner, how activities flow from the beginning to the end of a process. This diagram uses standard symbols (such as ovals for start and finish, rectangles for activities, diamonds for decisions) connected by arrows to indicate the direction of flow (Barosz et al. 2020). These symbols allow you to intuitively visualize each stage of the process and the relationships between them (NSB et al. 2022).

Operation flowcharts are fundamental tools in the industry for several reasons (Cazacu et al. 2021):

1. Process analysis and improvement: They allow the identification of bottlenecks, redundant activities, or inefficiencies in existing processes. By visualizing the complete workflow, areas of opportunity can be detected to optimize production and reduce costs.
2. Effective communication: Facilitates an understanding of the processes, for both personnel involved and those unfamiliar with them. A clear and concise diagram avoids misunderstandings and ensures that everyone is aligned in terms of objectives and tasks to be performed.
3. Process documentation: This serves as a historical record of how a process is carried out at a given point in time. It is useful for audits, training new employees, or implementing changes in the future.
4. Planning and control: They help establish a detailed work plan and monitor a process's progress. By visualizing the different stages, checkpoints can be identified, and key performance indicators (KPIs) can be established to evaluate performance.
5. Simulation and "what if" analysis: Operation flowcharts can be used to create simulation models that evaluate the impact of different scenarios or changes in the process. This is especially useful for strategic decision-making.

The benefits of the process flow diagram are as follows:

1. Clear visualization of processes, they are intuitive at each stage of the process, facilitating understanding by all involved, even those without deep technical knowledge (Duc et al. 2022).
2. Identifying bottlenecks by visualizing the workflow makes it possible to easily detect the stages that generate delays or backlogs, allowing corrective actions to optimize the workflow.
3. Elimination of redundant or unnecessarily repeated activities, which simplifies the process and reduces costs (Duc et al. 2022).
4. Improved communication, as it serves as a common language for all team members, ensuring that everyone understands responsibilities and expectations (Santosa and Mulyana 2023).
5. Facilitation of clear and concise documentation of processes, which is critical for training new employees, conducting audits, and managing change (Hernández-Soto et al. 2021).
6. They are the basis for automation because they are the first step, as they allow for identifying tasks that can be performed by computer systems.

7. What-if analysis, which can be obtained by simulating different scenarios in the diagram, allows evaluating the consequences of changes in the process and making more informed decisions (Duc et al. 2022).
8. Establishment of appropriate KPIs to measure process efficiency (Santosa and Mulyana 2023).
9. Quality improvement since it allows visualizing the workflow and possible sources of error can be identified, and measures can be taken to prevent defects in products or services (Hernández-Soto et al. 2021).
10. Cost reduction through optimizing processes, eliminating redundant activities, and improving overall efficiency leads to significant cost reductions (Santosa and Mulyana 2023).

However, the process flow diagram has the following disadvantages that are hardly criticized:

1. Oversimplification of reality, omitting important details that could affect the performance of the process. This can lead to an incomplete view and erroneous decisions based on partial information.
2. Difficulty in modeling complex processes or processes with multiple variables that may be difficult to represent clearly and concisely in a flowchart. This can limit its usefulness in some cases and make it difficult to represent.
3. Lack of updating in modern times, as processes change, and if flowcharts are not regularly updated to reflect changes, they can quickly become obsolete and lose their value (Santosa and Mulyana 2023).
4. Subjectivity in interpretation, as different people may understand different things when they look at it, which can lead to confusion and misunderstandings in production lines and their interactions (Valier 2020).
5. The time and resources invested in creating and maintaining flowcharts can be too much. In some cases, the cost of developing a detailed diagram may not be justified by the benefits obtained.
6. Resistance to change when areas for improvement are identified that employees do not accept since they are used to working with traditional methods (Valier 2020).
7. Flowcharts focus mainly on the structure of the process but do not always capture the behavior of the people involved in the process, leading to misrepresentations. This can limit their ability to identify problems related to motivation or communication (Santosa and Mulyana 2023).
8. Highly dynamic processes or processes with a high degree of uncertainty may be difficult to represent statically in a flowchart or change rapidly (Valier 2020).

An example of the process flow diagram is presented in Table 2.3.

TABLE 2.3 Example of a process flow diagram

LOCATION: HERNÁNDEZ AGENCY	*SUMMARY*			
ACTIVITY: DATA COLLECTION IN MORNING SHIFT	*EVENT*	*PRESENT*	*PROPOSED*	*SAVINGS*
Date: 05/01/2024	Operation	0		
Operator: P.L. Analyst: A.R.	Transportation	6		
Circle the appropriate method and type.	Delays	0		
Method: Present Proposed	Inspection	6		
Type: Worker Material Machine	Storage	2		
Comments:	Time (min)			
	Distance (ft)	217		
	Cost			
Description of events	Symbol	Time (min)	Distance (ft)	Reco-mmenda-tions
Collection of material in supply	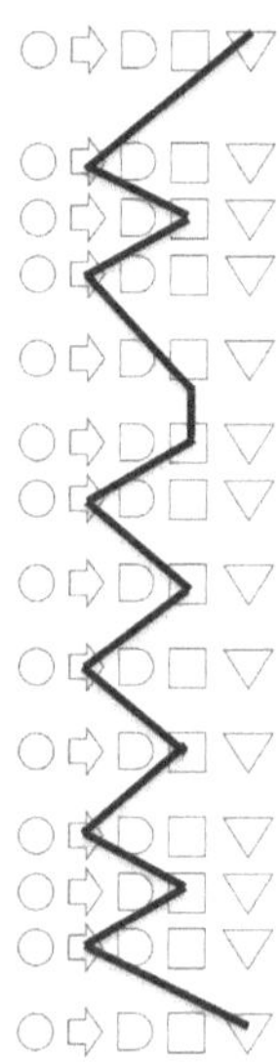			
To well 1			100	
Wagon inspection				
Toward well 2			20	
Water level inspection				
Steam inspection				
To well 3			25	
Vapor pressure inspection				
Toward well 4			20	
Inspection of energy consumption				
To well 5			32	
Steam inspection				
Toward supply			20	
Data capture to the server				

Source: Adapted from Bello Parra et al. (2020)

Therefore, a process flow diagram is crucial for assessing industrial operations. It provides a distinct visual depiction of each stage of the process, facilitating the identification of possible areas for enhancement and ensuring production consistency. Additionally, it functions as a reference resource for subsequent audits and staff training while enhancing communication across the teams engaged in the process. The process flow diagram is an effective tool for process validation, as it improves production efficiency and quality, assuring compliance with regulatory criteria. The integrity and security of products and processes in the sector depend on their execution.

2.2.4 Pareto Diagram

A Pareto diagram is a graphical tool used in quality management and continuous improvement to identify and prioritize problems or causes of a phenomenon in an industry (Garcia-Bernabeu et al. 2019). It is named after the Italian economist Vilfredo Pareto, who observed that 80% of the wealth in Italy was held by 20% of the population, which led to the development of the Pareto principle, also known as the 80/20 rule. This rule suggests that, in many cases, 80% of the effects come from 20% of the causes. In a Pareto diagram, a set of categorical data is represented in a bar chart. The bars represent frequencies or magnitudes of a series of events or problems, ordered from highest to lowest (Vallejo-Castillo et al. 2020). In addition, a line showing the cumulative percentage of these frequencies is included.

The practical application of a Pareto diagram in industry is extensive. For example, in a manufacturing environment, it could be used to identify defects in products, where it lists the types of defects and their frequency to determine which ones are most common (Armijal et al. 2023). This allows teams to focus their efforts on the areas that will have the greatest impact on quality improvement, thus maximizing the efficient use of resources and time. In addition, Pareto diagrams are valuable in project management to prioritize tasks, in failure analysis to identify root causes, and in customer service to improve customer satisfaction by addressing the most common problems first (Rodrigues dos Santos et al. 2018).

The Pareto diagram offers several advantages, as outlined in the following:

1. The Pareto diagram allows for quickly identifying the problems that generate the greatest impact on a process, facilitating decision-making and resource allocation (Bertocci et al. 2021).
2. Concentrating on the few causes that generate the most effects optimizes efforts and avoids unnecessary dispersion (Pacana and Czerwińska 2021).

3. It is a fundamental tool in continuous improvement processes, as it helps to visualize progress and establish realistic goals (Jaqin et al. 2020).
4. Pareto diagrams are easy to understand and communicate, facilitating all team members' participation in problem-solving (Bertocci et al. 2021).
5. By addressing the root causes of major problems, costs associated with defects, rework, and lost time are reduced (Pacana and Czerwińska 2021).
6. By eliminating or minimizing the causes of problems, process efficiency and productivity are enhanced.
7. Focusing on the problems that most affect the product or service quality significantly improves customer satisfaction (Pulido-Rojano et al. 2020).
8. The Pareto diagram provides an objective basis for decision-making, thereby avoiding decisions based on intuition or assumptions (Cedeño et al. 2018).
9. In addition to identifying problems, the Pareto diagram can also reveal opportunities for process improvement.
10. The Pareto diagram is a versatile tool that can be applied to various situations, from production to services.

While the Pareto diagram offers several benefits, it also has several drawbacks, as stated below:

1. The Pareto diagram offers a simplified view of reality by grouping data into categories. This can lead to losing sight of important details or interactions between factors that are not easily classified (Elena and Gerasimova 2019).
2. Classifying data can be subjective and vary according to the person conducting the analysis. This can lead to different diagram interpretations and make consensus difficult (Knop and Ziora 2022).
3. The Pareto diagram shows a correlation between categories but does not establish a causal relationship. That is, it does not indicate the root cause of the problems, only their relative frequency (Deidda 2010).
4. The Pareto diagram is based on historical data. If conditions change, it may no longer represent the current situation (Elena and Gerasimova 2019).
5. It is more suitable for discrete or categorical data. When working with continuous data, it may be necessary to group them into intervals, which may introduce some arbitrariness in the analysis (Deidda 2010).

6. The Pareto chart only considers the frequency of occurrence of problems without considering their severity or impact. An infrequent problem with serious consequences may be underestimated (Elena and Gerasimova 2019).
7. If not used correctly, the Pareto diagram can lead to erroneous conclusions. For example, if a too short time period is selected, the diagram may not accurately reflect the actual situation (Knop and Ziora 2022).
8. The Pareto diagram is a useful tool, but it is not a single solution to all problems. It should be complemented by other analysis tools to obtain a more complete picture of the situation (Knop and Ziora 2022).

Figure 2.2 illustrates a Pareto diagram utilized in the production of automobile components. The Pareto graphic demonstrates that the principal sources of failures in the manufacturing line are nonstandardized operations, inadequately trained staff, and unclear visual aids. The remaining reasons contribute minimally; however, these three causes constitute 80% of the manufacturing delays.

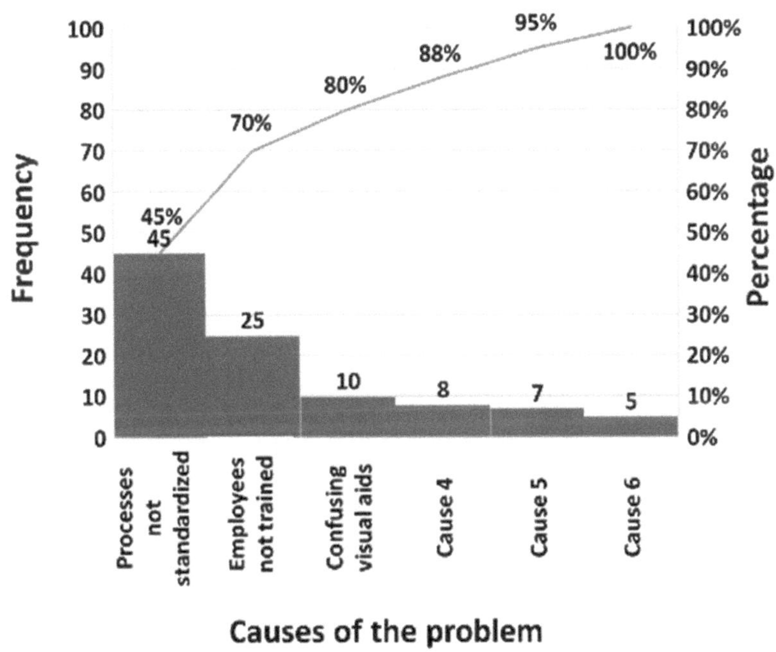

FIGURE 2.2 Example of Pareto diagram

Consequently, it is deducted that the most substantial decrease in delays and enhancements in the overall efficiency of the production process would arise from focusing efforts on optimizing the supply of components and the efficacy of the assembly machinery. The application of the Pareto diagram in process validation aids in identifying and prioritizing the most crucial causes or concerns, facilitating effective resource allocation and a strategic focus on essential areas to achieve substantial improvements (Bertocci et al. 2021). In summary, the Pareto diagram is an efficient instrument for discerning the most critical issues or causes and prioritizing subsequent actions (Knop and Ziora 2022).

In conclusion, the Pareto diagram is an effective instrument for process validation. It offers a transparent, visual method to discern and prioritize the most critical issues or obstacles within a process, enabling validation teams to concentrate on areas that influence process quality and efficiency. Utilizing this technology proficiently enables organizations to enhance their processes and satisfy the requisite quality standards. The Pareto diagram is essential for continuous improvement and operational excellence.

2.2.5 Ishikawa Diagram

The Ishikawa diagram, commonly referred to as the Fishbone diagram or cause-and-effect diagram, serves as a quality tool for identifying potential causes of a problem or undesirable effect (Sitko et al. 2021). The Ishikawa diagram has been widely used in various sectors and industries, including manufacturing, project management, healthcare, and environmental management (Schmitt and Burchinal 2011). This diagram allows different potential causes to be visually organized into main categories and subcategories, which facilitates the analysis and understanding of the relationships and connections between them (Florea et al. 2022).

Once the problem is identified, a working team composed of representatives from different areas related to the problem is assembled (Barsalou 2023). This team discussed and analyzed the possible causes of this problem. During the brainstorming session, several possible causes were identified and grouped into six categories: labor, machines, methods, materials, measures, and the environment (Anshori and Karya 2022). These categories represent different areas that can influence assembly line production. The team uses the Ishikawa diagram to visualize these possible causes and their relationship to the main problem (Rodríguez-Palomino 2021). The head of the fish represents the main problem, and the spines or branches represent various categories of possible causes (Sitko et al. 2021). When using the Ishikawa diagram, a horizontal line is drawn in the center of the page, starting from the head of the fish, on which the problem or undesired effect is labeled (Bouaoudate et al. 2018).

From there, diagonal lines are drawn resembling the backbone of a fish, and on these lines, different potential causes related to the problem are written (Lipińska et al. 2019). These causes are classified into main categories, including people, processes, materials, environment, and methods (Lipińska et al. 2019). Each of these main categories may have additional subcategories for greater specificity in identifying causes. Each of these main categories may have additional subcategories for more specificity in identifying causes. The Ishikawa diagram is a valuable tool for analyzing and understanding the possible causes of a problem or undesired effect, as it helps to visualize all possible influences and connections that may contribute to the problem in a clear and structured manner (Bouaoudate et al. 2018). In addition, the Ishikawa diagram helps prioritize the identified causes and make informed decisions on what actions should be taken to address the problem at hand (Worsfold et al. 2019). By identifying and understanding the root causes of a problem, it is possible to implement effective solutions to prevent its recurrence in the future. The Ishikawa diagram has several advantages that make it an effective tool for identifying and analyzing the causes of a problem (Realyvásquez-Vargas et al. 2020). Some advantages of the Ishikawa diagram are as follows:

1. During a brainstorming session, the Ishikawa diagram can be easily adapted (Ortiz-Hernandez et al. 2019). This allows participants to focus on the content of the problem and facilitates data collection.
2. The structure of the Ishikawa diagram stimulates group participation and helps maintain focus on the content of the problem (Realyvásquez-Vargas et al. 2020).
3. Using the Ishikawa diagram in practice, the basic groups of causes responsible for quality problems can be determined (Sitko et al. 2021).
4. The Ishikawa diagram encourages creative thinking process and teamwork by allowing multiple people to contribute to identifying possible causes and relevant categories (Kreager et al. 2022).

These advantages demonstrate that the Ishikawa diagram is a powerful, universal tool applicable to quality management, which can be used to identify the causes of various quality problems, specify the most important reasons, improve efficiency, and reduce problems in production processes (Morell-Santandreu et al. 2021). In summary, the Ishikawa diagram offers several advantages, such as its adaptability in brainstorming sessions and its ability to identify the root causes of quality problems.

For all its advantages, the Ishikawa diagram also has some disadvantages. First, there can be a tendency to rely solely on brainstorming to identify problem causes, which can limit the diversity of perspectives and potential solutions (Talancon et al. 2020). Another disadvantage is that the Ishikawa diagram may be insufficient for identifying the root cause of the problem (Knop 2021). This is because the diagram only provides hypotheses about possible causes, and further analysis and testing are required to confirm the root cause (Anshori and Karya 2022). In addition, the process of constructing a diagram can be a complex process that requires additional time and resources (Talancon et al. 2020). These disadvantages can make it difficult to implement it in organizations with limited resources or situations where a rapid response is required (Barsalou 2023).

The following is an example of an Ishikawa diagram used to identify the possible causes of a high number of defects in an automotive component production line. In this case, the problem identified was the high number of defects in an automotive component production line. The objective was to identify the possible causes of the defects and take corrective actions to improve the efficiency of the production process and increase the quality of the components. Examples of the possible causes within each category are shown in Figure 2.3. Ishikawa diagram analysis allows the work team to identify the possible causes of the high number of defects and prioritize the necessary corrective actions to improve process efficiency and increase quality.

Thus, the Ishikawa diagram is an invaluable tool for process validation. Using this technique, companies can effectively identify the potential causes of problems in the validation process. By understanding these root causes, organizations can take corrective and preventive actions to improve the efficiency and quality of their processes. By integrating the Ishikawa diagram into process validation, companies can improve their ability to meet regulatory standards and satisfy customers' needs and expectations.

2.2.6 Design of Experiments

DOE is a methodology used in research to test hypotheses and draw reliable conclusions (Casler 2015). It consists of carefully planning how an experiment will be conducted, with the goal of maximizing the information that can be obtained from the fewest trials (Pro Academic Writers 2024). There are different types of experimental designs, such as factorial, completely randomized, and randomized block designs (Pro Academic Writers 2024). Each of these designs has its own advantages and disadvantages; therefore, it is important to select the most appropriate one for the type of experiment to be conducted (Collins et al. 2009). In addition, DOE includes random assignment

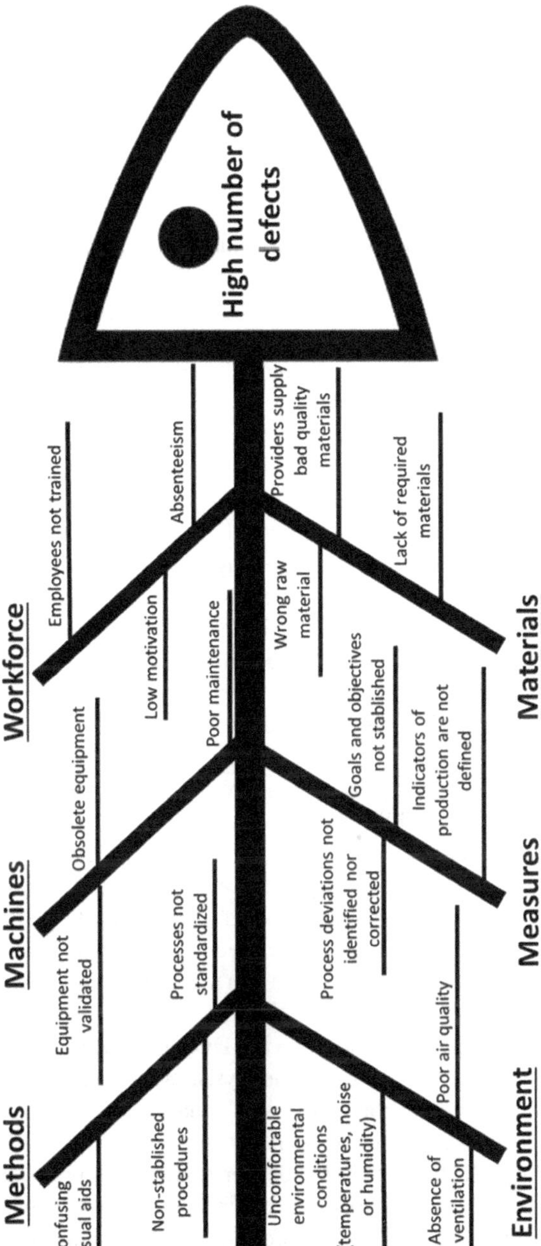

FIGURE 2.3 Example of an Ishikawa diagram for the problem of defects in an automotive component production line

of treatments, replication of tests, and control of variables that may affect the results (Hahn 1984). This was performed to reduce bias and ensure the validity of the results obtained.

DOE not only is fundamental in scientific research but also plays a crucial role in the industry (Dorta 2022). In the industry, DOE is used to optimize processes, improve product quality, reduce costs, and make data-driven decisions (Tripathi et al. 2021). One of the most common applications of DOE in industry is the optimization of manufacturing processes. By applying different experimental designs, companies can identify the optimal configuration of variables to maximize efficiency and minimize waste (Sadiq et al. 2021). This can lead to significant savings in operating costs and substantial improvements in final product quality (Fan, Leu, and Krischke 2016). Another important application of DOE in the industry is quality improvement. By conducting controlled and systematic tests, companies can identify factors that influence product quality and take steps to optimize them (Montgomery 2001). This can translate into reduced defects, increased customer satisfaction, and increased competitiveness in the market (Mathiyazhagan et al. 2022).

DOE is a powerful tool that can be pivotal in process validation within an industry (Sahu and Andhare 2018). By applying an experimental design, companies can identify and quantify the key factors that influence the performance and quality of their production processes (Delgado Fernandez 2020).

An example of the application of DOE in the industry is its use to optimize a fuse's resistance and capacitance ratings. In a study by Torres Sosa et al. (2020), an experimental scheme for optimizing fuse manufacturing process parameters was presented. The results showed that, through the analysis of the capacity indexes, it was possible to increase the process capacity, going from a Cpk value of 0.25 to 1.85 (Tanco et al. 2009). This indicates a significant improvement in quality and compliance with the product specifications (Torres Sosa et al. 2020). Another example of a DOE application in the industry is the improvement in the strength and shelf life of a protective film. In this case, experimental design techniques were used to optimize the parameters of the manufacturing process, which resulted in a substantial improvement in the product's shelf life (Antony et al. 2012).

There are numerous benefits to applying DOE to industrial process validation. By applying rigorous statistical analysis, companies can effectively identify the factors affecting their processes' quality and performance and make informed decisions to improve them (Torres Sosa et al. 2020). In addition, experimental design allows for a deeper understanding of the interactions between variables and factors that can influence process performance (Mathiyazhagan et al. 2022). This is critical for the continuous optimization and identification of solutions that significantly improve efficiency and product quality.

The three tools that are widely used in DOE and are valuable for process validation are correlation and regression analyses (Sauer et al. 2006; X. Li and Yang 2020) and ANOVA (Jankovic et al. 2021). It is necessary to collect data on the variables under study to conduct such analyses. After collecting the necessary data, it is important to perform a detailed correlation and regression analysis to better understand the relationships between variables (Tripathi et al. 2021). In the case of correlation analysis, this is a statistical tool that allows one to assess whether a relationship exists between two variables and to measure the strength and direction of that relationship (Dixit et al. 2022) through the numerical value of the correlation coefficient (Mathiyazhagan et al. 2022). The most commonly used correlation coefficients are Pearson's correlation coefficient for continuous variables and Spearman's correlation coefficient for ordinal variables (Dixit et al. 2022).

There are different types of correlations, such as positive correlation, where both variables increase together, and negative correlation, where one variable decreases as the other increases. By understanding the correlation between variables, patterns, and trends in the data can be identified, which is critical for informed decision-making (Alhuraish et al. 2017). For example, correlation analysis can help uncover whether there is a relationship between increased advertising spending and increased sales, which is critical in determining business strategies. In correlation analysis, it is also essential to remember that the identified relationship does not necessarily imply a causal relationship. In other words, just because two variables are correlated, this does not mean that one directly causes a change in the other (Alhuraish et al. 2017). This is a key point to consider when interpreting the correlation results.

On the other hand, regression analysis is a statistical technique that allows us to understand the relationship between a dependent variable and one or more independent variables (Tripathi et al. 2021). Through regression analysis, it is possible to model the relationship between these variables and predict the dependent variable's future values based on the independent variables' values (Dixit et al. 2022). In other words, regression helps predict the value of one variable as a function of another variable (Dixit et al. 2022).

There are different types of regression, such as linear regression, which assumes a linear relationship between variables, and nonlinear regression, which allows more complex relationships to be modeled. Linear regression is widely used due to its simplicity and ease of interpreting results; however, in many cases, the relationship between variables may not be strictly linear, which justifies using other types of regression (Gopi and Saleeshya 2022). When conducting a regression analysis, it is critical to consider the model's fit, the coefficients' significance, and the interpretation of the terms in the regression equation (Dixit et al. 2022). In addition, it is essential to assess the model's goodness of fit using measures such as the coefficient of determination and standard error of the estimation (Sadiq et al. 2021).

By using regression, one can not only understand the relationship between variables but also identify the relative contribution of each independent variable to predicting the dependent variable (Dixit et al. 2022). This is crucial for making informed decisions, as it allows one to assess which factors significantly impact the outcome one is trying to predict.

In addition, it is essential to remember that regression, such as correlation, does not imply causation. That is, because an independent variable is related to the dependent variable, it does not mean that the former directly causes changes in the latter (Gopi and Saleeshya 2022). Regression provides information about the association between variables, but causal inference should be made with caution, and other factors that may influence the relationship should be considered (Dixit et al. 2022). By delving deeper into correlation and regression analysis, valuable information can be obtained about the behavior of the variables under study, which can provide a solid basis for informed decision-making (Tripathi et al. 2021).

Correlation and regression analyses play a key role in process validation. By applying these statistical techniques in manufacturing, critical information can be obtained to understand the relationships among the variables involved in the process (Tripathi et al. 2021). In the manufacturing environment, correlation and regression can help identify the variables that significantly impact the quality of the final product (Alhuraish et al. 2017). For example, by studying the relationship between manufacturing temperature and the physical properties of the product, such as strength, durability, or toxicity, optimal temperature conditions can be determined to achieve the desired quality standards. In addition, regression analysis can predict future values of certain variables as a function of others, which is crucial for planning manufacturing processes (Tripathi et al. 2021). For example, by understanding how variations in production speed affect the efficiency and uniformity of the final product, adjustments can be made to optimize process performance.

Correlation and regression analyses have several advantages. For example, by performing a detailed correlation analysis, one can identify the presence of associations between variables and assess the strength and direction of these relationships (Tripathi et al. 2021). This in-depth understanding of correlation can provide valuable information for decision-making, ranging from business strategies to policy and program development (Mathiyazhagan et al. 2022). However, by exploring regression at greater depths, one can appreciate the complexities of modeling the relationship between variables (Gopi and Saleeshya 2022). The ability to predict the behavior of one variable as a function of other variables offers a significant advantage in strategic planning and outcome forecasting (Dixit et al. 2022).

However, it is important to recognize that correlation and regression have limitations. For example, neither technique implies causality, which means that the association identified between variables does not necessarily indicate a cause-and-effect relationship (Dixit et al. 2022), which requires careful analysis and interpretation of the results. In addition, both techniques can be sensitive to outliers and underlying assumptions, underscoring the importance of critical evaluation and consideration of multiple factors when interpreting results (Mathiyazhagan et al. 2022). Despite these limitations, correlation and regression analyses remain invaluable tools for understanding the relationships between variables and predicting behavior in various contexts (Alhuraish et al. 2017). A detailed understanding of the advantages and disadvantages of these statistical techniques is essential for their effective application in decision-making and manufacturing process validation (Tripathi et al. 2021).

ANOVA is a statistical technique used to determine if there are significant differences between the means of three or more independent groups (Abdullahi et al. 2019). This tool is essential in the industry for comparing different processes, materials, or production conditions to make informed decisions based on data (Barczyk-Pawelec et al. 2022). The ANOVA is based on the decomposition of the total variability observed in a data set into components attributable to different sources of variation. In simple terms, it assesses the relationship between the variability "within" each group and the variability "between" the groups (Tian et al. 2020). If the between-group variability is considerably greater than the within-group variability, the differences between means are likely to be statistically significant (Abdullahi et al. 2019). To successfully apply an ANOVA in industry, specific requirements or assumptions must be met (Vencúrik et al. 2021):

1. Normality: The data in each group are assumed to follow a normal distribution. Although ANOVA is robust to deviations from normality, especially with large sample sizes, it is important to verify this assumption using normality tests or plots such as histograms or Q-Q plots.
2. Homogeneity of variances: Also known as homoscedasticity, this assumption implies that the variances of the different groups should be approximately equal. Violation of this assumption can lead to incorrect conclusions. Tests such as Levene's or Bartlett's can be used to evaluate this assumption.
3. Independence: The observations must be independent of one another. This means that the result of one observation should not influence the results of the others. This assumption is critical since its violation may bias the results.
4. Interval or ratio scale: The data must be measured on an interval or ratio scale. This ensures that differences between values are significant and comparable.

Applying ANOVA in the industry allows engineers and quality professionals to identify which factors or conditions significantly impact the process or product, thereby facilitating process optimization, reducing variability, and promoting continuous improvement. For example, in quality control, ANOVA can help determine whether different machines, work shifts, or raw material batches significantly affect final product characteristics.

ANOVA offers the following advantages:

1. ANOVA allows us to identify which factors or variables have a significant impact on a process or product. This is crucial in industry, where multiple variables can influence the result. By identifying the most important factors, companies can focus their improvement efforts on areas that truly affect quality or efficiency (Nagamine et al. 2009).
2. ANOVA helps optimize industrial processes by determining which conditions or process configurations yield the best results (Faria et al. 2018). This can lead to cost reduction, as waste is minimized and operational efficiency is maximized. For example, in manufacturing, ANOVA can be used to find the optimal combination of temperature, pressure, and processing time.
3. ANOVA is a valuable tool for the continuous improvement cycle, as it allows process changes to be evaluated and their effectiveness to be measured. By conducting periodic tests, companies can ensure that implemented improvements contribute to quality and efficiency, and adjust strategies as necessary (Cao et al. 2019).
4. ANOVA provides a solid statistical basis for decision-making. Instead of relying on assumptions or intuition, engineers and managers can base decisions on quantitative analyses that demonstrate the effectiveness of different strategies or process changes (Gabriel et al. 2016).
5. One of the primary objectives in the industry is to minimize process variability, thereby ensuring product consistency. ANOVA helps to identify sources of variability and implement measures to control them, resulting in more uniform and higher-quality products (Wang et al. 2022).

However, ANOVA has several disadvantages:

1. ANOVA requires certain assumptions to be met, including normality of the data, homogeneity of variances, and independence of observations. If these assumptions are not met, the results may be unreliable or misleading. In industrial practice, data often do not perfectly meet these assumptions, which can limit the applicability of ANOVA (Stepanek et al. 2022).

2. ANOVA is sensitive to the presence of outliers, which can distort the results and lead to incorrect conclusions (Yirga et al. 2020). In industrial settings, where data may be noisy or contain measurement errors, outliers are common and can significantly affect the analysis.
3. Although ANOVA can indicate whether there are significant differences between groups, it does not specify which groups are different from each other. This requires additional post hoc analyses, such as Tukey or Bonferroni tests, which complicate interpretation and increase the time and effort needed to draw clear conclusions (Stepanek et al. 2022).
4. ANOVA is the most effective when used in well-structured experimental designs. However, in industry, we often work with observational or nonexperimental data, which can limit the ability of ANOVA to identify causal relationships. This can lead to decisions based on correlations rather than causalities.
5. The correct application and interpretation of ANOVA requires a considerable level of statistical expertise. In summary, although ANOVA is a powerful tool for data analysis in the industry, its disadvantages include the need to meet strict assumptions, its sensitivity to outliers, complexity in interpreting results, limitations in experimental design, and the need for advanced statistical knowledge. These disadvantages should be considered when deciding whether ANOVA is the appropriate tool for a specific analysis.

For example, a company that manufactures LED lamps is interested in improving the consistency of brightness in their products. The company experimented to investigate how certain manufacturing process parameters affect the lamps' brightness. Twenty lamps were randomly selected from daily production to measure the following variables:

Dependent variable: Lamp brightness (measured in lumens).
Independent variables:

- Operating current intensity (IC) (mA).
- Exposure time (ET) to elevated temperatures during the curing process (min).

Table 2.4 presents the collected data.

To carry out the correlation analysis, the correlation coefficient $r_{x\text{-}y}$ is calculated to determine the Pearson correlation ($-1 \leq r \leq 1$) between lamp

TABLE 2.4 Data for DOE example

LAMP	*CURRENT INTENSITY (MA)*	*EXPOSURE TIME (MIN)*	*LUMINOSITY (LUMENS)*
1	150	10	380
2	180	12	410
3	160	8	390
4	140	9	370
5	170	11	400
6	155	10	385
7	175	12	415
8	145	8	375
9	165	9	395
10	150	11	385
11	180	10	410
12	160	12	400
13	140	8	370
14	170	9	390
15	155	11	395
16	175	10	405
17	145	12	380
18	165	8	370
19	152	9	380
20	178	11	410

brightness and each independent variable (CI and ET). This is done by applying Equation (2.1)

$$r_{x-y} = \frac{\operatorname{cov}(x,y)}{S_x S_y} = \frac{n\sum xy - \sum x \sum y}{\sqrt{\left[n\sum x^2 - \left(\sum x\right)^2\right]\left[n\sum y^2 - \left(\sum y\right)^2\right]}} \tag{2.1}$$

where x is the independent variable (IC and ET); y is the dependent variable (luminosity); Cov (x, y) is the covariance between x and y; S_x is the standard deviation of x; S_y is the standard deviation of y; and n is the sample size (20). Thus, a correlation coefficient of 0.8583 was obtained for the IC—luminosity relationship, and a correlation coefficient of 0.6324 was obtained for the ET–luminosity relationship. These correlation coefficients indicate a strong positive correlation between independent and dependent variables.

To perform the regression analysis, a multiple linear regression model was fitted to predict the lamp brightness as a function of the IC and ET. Equation (2.2) is a multiple linear regression equation:

$$y = \beta_0 + \beta_1 x_1 + \beta_2 x_2 + \varepsilon \tag{2.2}$$

where β_0, β_1, and β_2 are the regression coefficients and ε is the error term. The coefficients β_0, β_1, and β_2 are calculated using statistical methods such as the least squares method. For the data in this example, the results were β_0 =218.93, β_1 = 1.83, and β_2 = 3.08. Therefore, the multiple line regression equation is

$$Brightness = 218.93 + 1.83 \times CI + 3.08 \times ET$$

This equation allows the prediction of the expected brightness of an LED lamp based on the IC and ET values during the manufacturing process.

Next, ANOVA was performed to determine whether the independent variables (IC and ET) significantly affected lamp brightness. First, null hypothesis H_0 and alternative hypothesis H_1 are formulated as follows:

H_0: The IC and ET did not significantly affect the lamp brightness.
H_1: The IC and ET significantly affect the lamp brightness.

Next, the parameters shown in Equations (2.3)–(2.8) were calculated (Tables 2.5 and 2.6).

$$SS\ total = \sum y^2 - \frac{\sum (Y)^2}{n} \tag{2.3}$$

$$SST = \sum \left(\frac{T_C^2}{n_C} \right) - \frac{\sum (Y)^2}{n} \tag{2.4}$$

$$SSE = SS\ total - SST \tag{2.5}$$

$$V_1 = K - 1 \tag{2.6}$$

$$V_2 = n - K \tag{2.7}$$

$$\text{Number of tests} = \frac{K(K-1)}{2} \tag{2.8}$$

where T_C is the sum of squares due to treatment (sample), N_C is the number of observations for each treatment (sample), K is the number of treatments

TABLE 2.5 ANOVA parameters

SOURCE OF VARIATION	*SS*	*DF*	*MS*	*F*	*P-VALUE*
Treatments	*SST*	*K-1*	*TSS/(K-1)*	*MST/MSE*	
Error	*SSE*	*n-K*	*SSE/(n-K)*	*MST/MSE*	
Total	Total *SS*	n-1			

**SS* = sum of squares, df = degrees of freedom, *MS* = mean of squares

TABLE 2.6 ANOVA results

SOURCE OF VARIATION	*SS*	*DF*	*MS*	*F*	*P-VALUE*
Regression	2295.55	2	1147.78	56.72	<0.05
Error	540	17	31.76		
Total	2835.55	19			

(samples), *SSE* is the sum of squares of error, V_1 is the degrees of freedom in the numerator, V_2 is the degrees of freedom in the denominator, *MST* is the mean squares of treatment, *MSE* is the mean squares error, and *F* is the distribution of test statistic.

Table 2.6 shows the results of ANOVA.

To calculate the *F*-statistic and its corresponding *p*-value, Equation (2.9) was used:

$$F = \frac{MST}{MSE} \tag{2.9}$$

where MST and MSE are the mean square of the regression and error, respectively. In this case:

$$F = \frac{1147.68}{31.76} = 36.10$$

The *p*-value associated with *F* was less than 0.05, indicating that the null hypothesis was rejected. It was concluded that at least one of the predictors (IC or ET) significantly affected lamp brightness. According to the correlation analysis, the IC and ET strongly correlate with lamp brightness. Furthermore, the regression analysis showed that both parameters were significant predictors of brightness.

Process validation in industry is critical for ensuring the reliability and consistency of product manufacturing (N. Zhou et al. 2020). The experimental design provides the ability to perform systematic and controlled tests that validate the effectiveness of operating procedures and parameters, thereby contributing to informed decision-making and the continuous improvement of industrial processes. In summary, experimental design is a powerful tool that can significantly improve the validation process. Its application can lead to the identification of improvements, cost savings, and optimization of final product quality. Therefore, it is strongly recommended that this design be considered and used for process validation.

In addition, correlation and regression are fundamental statistical tools that allow an understanding of the relationships between variables and predict the behavior of one variable as a function of the other. By using these techniques carefully and critically, valuable information can be obtained for informed decision-making in different areas, ranging from business to scientific research. In this way, correlation and regression analyses provide a solid basis for decision-making based on the validation of manufacturing processes, allowing the identification of critical variables, predicting future behavior, and optimizing the quality and efficiency of production.

Regarding ANOVA, it is concluded that it is a powerful tool in process validation, allowing the comparison of means across several samples and determining whether there are significant differences between them. This is crucial in the industry to ensure the quality and consistency of the manufacturing processes. Using ANOVA, companies can identify and correct process problems, leading to greater efficiency and cost reduction. Thus, ANOVA is a valuable tool that can provide crucial information for process validation in various industries.

2.3 CONCLUSION

After presenting the theoretical foundations of various tools for process validation, it was concluded that process validation is essential to ensure quality and consistency in production. Auxiliary tools such as 8Ds, flowcharts, Pareto diagrams, and DOE can provide a clearer view of the problems and challenges in the validation process. By effectively integrating these tools, organizations can identify and address areas for improvement with greater accuracy, ultimately leading to enhanced efficiency and improved production quality. Implementing ancillary tools for process validation offers a significant advantage by providing a deeper understanding of potential problems and areas for improvement. In the specific case of the structured 8Ds approach, challenges are addressed systematically, while process flow diagrams provide a clear visual representation of the steps involved. On the other hand, DOE allows the effective identification of factors that affect the quality and performance of processes and thus makes decisions to improve them.

In addition, the use of Pareto and Ishikawa diagrams effectively identifies the root causes of problems, which in turn allows a more precise intervention to address them. The application of DOE techniques and ANOVA allows the effect of multiple variables on the process to be evaluated, contributing to a more holistic approach to process validation.

By effectively integrating these tools, organizations can more accurately identify areas for improvement, anticipate potential deviations, and optimize the efficiency of the validation process. The combination of these tools provides a solid foundation for informed decision-making and continuous improvement in production quality and consistency.

REFERENCES

Abdullahi, Said Rajab, Muhammed Raşit Öner, and Osman Nuri Ata. 2019. "Fractional Factorial Design Application for the Determination of Parameters Affecting KOH and HCl Generation from Simulated Wastewater Solution by Bipolar Membrane Electrodialysis." *European Journal of Science and Technology* 17: 866–73. https://doi.org/10.31590/ejosat.646850.

Alhuraish, Ibrahim, Christian Robledo, and Abdessamad Kobi. 2017. "A Comparative Exploration of Lean Manufacturing and Six Sigma in Terms of Their Critical Success Factors." *Journal of Cleaner Production* 164 (October): 325–37. https://doi.org/10.1016/J.JCLEPRO.2017.06.146.

Ando, Guiseppe. 2024. "The Dynamics to Build a Winning Team." https://www.peoplehum.com/blog/the-dynamics-to-build-a-winning-team.

Anju, Goyal, and Priyambada Pandey. 2017. "Process Validation of Pharmaceutical Dosages Form: A Review." *Biomedical Journal of Scientific & Technical Research* 1 (5): 1–9. https://doi.org/10.26717/BJSTR.2017.01.000463.

Anshori, Mohamad Yusak, and Denis Fidita Karya. 2022. "A Analysis of Factors Affecting Waste Using Value Stream Mapping (VSM) on Kaosbekas.Id." *Jurnal Entrepreneur Dan Entrepreneurship* 11 (2): 139–54. https://doi.org/10.37715/JEE.V11I2.2750.

Antony, Jiju, E.V. Gijo, and S.J. Childe. 2012. "Case Study in Six Sigma Methodology: Manufacturing Quality Improvement and Guidance for Managers." *Production Planning & Control* 23 (8): 624–40. https://doi.org/10.1080/09537287.2011.576404.

Armijal, W. A. Marlina, and R. A. Hadiguna. 2023. "The Evaluation of Supply Chain Risk Management on Smallholder Layer Farms." *IOP Conference Series: Earth and Environmental Science* 1182 (1): 012082. https://doi.org/10.1088/1755-1315/1182/1/012082.

ASTM International. 2019. "Standard Guide for Corrective Action and Preventive Action (CAPA) for the Cannabis Industry." https://www.astm.org/d8229-19.html.

Barczyk-pawelec, Katarzyna, Krystian Rubajczyk, Małgorzata Stefańska, Łukasz Pawik, and Wioletta Dziubek. 2022. "Characteristics of Body Posture in the Sagittal Plane in 8–13-Year-Old Male Athletes Practicing Soccer." *Symmetry* 14 (2): 210. https://doi.org/10.3390/SYM14020210.

Barosz, Piotr, Grzegorz Gołda, and Adrian Kampa. 2020. "Efficiency Analysis of Manufacturing Line with Industrial Robots and Human Operators." *Applied Sciences* 10 (8): 2862. https://doi.org/10.3390/APP10082862.

Barsalou, Matthew. 2023. "Root Cause Analysis in Quality 4.0: A Scoping Review of Current State and Perspectives." *TEM Journal* 12: 73–79. https://doi.org/10.18421/TEM121-10.

Bello Parra, Daniel, Félix Murrieta Domínguez, and Carlos Alberto Cortes Herrera. 2020. "Nálisis de Tiempos y Movimientos En El Proceso de Producción de Vapor de Una Empresa Generadora de Energías Limpias." *Ciencia Administrativa* 2020 (1): 1–9.

Bertocci, Francesco, Andrea Grandoni, Monica Fidanza, and Rossella Berni. 2021. "A Guideline for Implementing a Robust Optimization of a Complex Multi-Stage Manufacturing Process." *Applied Sciences* 11 (4): 1418. https://doi.org/10.3390/APP11041418.

Bose, Tarun Kanti. 2012. "Application of Fishbone Analysis for Evaluating Supply Chain and Business Process—A Case Study on the St James Hospital." *International Journal of Managing Value and Supply Chains* 3 (2): 17–24. https://doi.org/10.5121/ijmvsc.2012.3202.

Bouaoudate, M., Jaouad El Harti, M. Maamar, and R. Abouqal. 2018. "Design of a Typical Residence for the Elderly in Morocco." *Journal of Public Health in Africa* 9 (3): 146–49. https://doi.org/10.4081/jphia.2018.835.

Bremmer, Jurgen. 2015. *Update and Upgrade the Current Quality Assurance in the Global Supply Chain of a Heavy Truck Manufacturing Company*. University of Twente.

Calvo-Manzano, Jose, Gonzalo Cuevas Agustin, Tomas San Feliu Gilabert, and Jose De J. Jimenez Puello. 2016. "DEPROVAL:Herramienta Para La Definición Del Proceso de Validación En Pequeñas Organizaciones de Software." In *Iberian Conference on Information Systems and Technologies, CISTI*, July, 1–6. Gran Canaria, Spain: IEEE Computer Society. https://doi.org/10.1109/CISTI.2016.7521500.

Cao, Wen, Shun Li Wang, Carlos Fernandez, Chuan Yun Zou, Chun Mei Yu, and Xiao Xia Li. 2019. "A Novel Adaptive State of Charge Estimation Method of Full Life Cycling Lithium-Ion Batteries Based on the Multiple Parameter Optimization." *Energy Science & Engineering* 7 (5): 1544–56. https://doi.org/10.1002/ESE3.362.

Carmignani, Gionata. 2009. "Modified QFD and Problem-Solving Techniques Integrated Approach Implementing Corrective Actions: A Case Study in an Italian Manufacturing Plant." *Quality and Reliability Engineering International* 25 (2): 241–52. https://doi.org/10.1002/QRE.967.

Casler, Michael D. 2015. "Fundamentals of Experimental Design: Guidelines for Designing Successful Experiments." *Agronomy Journal* 107 (2): 692–705. https://doi.org/10.2134/AGRONJ2013.0114.

Cazacu, Carmen Cristiana, Florina Chiscop, and Dragos Alexandru Cazacu. 2021. "Using IoT to Dynamically Test Smart Connected Devices." In *MATEC Web of Conferences 343. 10th International Conference on Manufacturing Science and Education—MSE 2021*, edited by N.F. Cofaru and M. Inţă. Vol. 343, 1–9. Sibiu, Romania: EDP Sciences. https://doi.org/10.1051/MATECCONF/202134303004.

Cedeño, Edilberto Antonio Llanes, José Bolívar Carguachi-Caizatoa, and Juan Carlos Rocha-Hoyos. 2018. "Energy and Exergy Evaluation in a 1.6L Otto Cycle Internal Combustion Engine." *Enfoque UTE* 9 (4): 221–32. https://doi.org/10.29019/ENFOQUEUTE.V9N4.365.

Collins, Linda M., John J. Dziak, and Runze Li. 2009. "Design of Experiments with Multiple Independent Variables: A Resource Management Perspective on Complete and Reduced Factorial Designs." *Psychological Methods* 14 (3): 202–24. https://doi.org/10.1037/A0015826.

Cuenca, Llanos, Ángel Ortiz, and Andrés Boza. 2008. "Desarrollo de Una Herramienta Software Para La Vista de Información de La Arquitectura CIMOSA." *Información Tecnológica* 9 (3): 97–106. https://doi.org/10.1612/inf.tecnol.3938it.07.

Deidda, R. 2010. "A Multiple Threshold Method for Fitting the Generalized Pareto Distribution to Rainfall Time Series." *Hydrology and Earth System Sciences* 14 (12): 2559–75. https://doi.org/10.5194/HESS-14-2559-2010.

Delgado Fernández, Mercedes. 2020. "Uso Del Diseño de Experimentos Para La Innovación Empresarial." *Revista de Métodos Cuantitativos Para La Economía y La Empresa* 29 (January): 38–56. https://doi.org/10.46661/revmetodoscuanteconempresa.2450.

Dixit, Aasheesh, Suresh Kumar Jakhar, and Patanjal Kumar. 2022. "Does Lean and Sustainable Manufacturing Lead to Industry 4.0 Adoption: The Mediating Role of Ambidextrous Innovation Capabilities." *Technological Forecasting and Social Change* 175 (February): 121328. https://doi.org/10.1016/J.TECHFORE.2021.121328.

Dixon, Andrew. 2022. "ITIL Problem Management." In *Practical Guide to IT Problem Management*, edited by Andrew Dixon. 1st ed., 51–56. New York: Auerbach Publications. https://doi.org/10.1201/9781003119975-11/ITIL-PROBLEM-MANAGEMENT-ANDREW-DIXON.

Dorta, Roberto. 2022. "Cómo Hacer Las Mejores Palomitas de Maíz." *Café y Teoremas*. https://elpais.com/ciencia/cafe-y-teoremas/2022-12-31/como-hacer-las-mejores-palomitas-de-maiz.html.

Duc, Minh Ly, Petr Bilik, and Thuy Duy Truong. 2022. "Design of Industrial System Using Digital Numerical Control." *Quality Innovation Prosperity* 26 (3): 135–50. https://doi.org/10.12776/QIP.V26I3.1747.

Elena, Bykowa, and Irina Gennadyevna Gerasimova. 2019. "Land Plot Selection Rationale for the Location of Linear Facilities." *Land* 8 (4): 67. https://doi.org/10.3390/LAND8040067.

Fan, Mei Niang, Jun Der Leu, and Andre Krischke. 2016. "The Evaluation of Green Manufacturing: A DEA-Based Approach." In *IEEE International Conference on Industrial Engineering and Engineering Management*, edited by Kadarsah Suryadi and Budi Hartono, December, 65–69. Bali, Indonesia: IEEE Computer Society. https://doi.org/10.1109/IEEM.2016.7797837.

Faria, Melissa, Inmaculada Fuertes, Eva Prats, Jose Luis Abad, Francesc Padrós, Cristian Gomez-Canela, Josefina Casas, et al. 2018. "Analysis of the Neurotoxic Effects of Neuropathic Organophosphorus Compounds in Adult Zebrafish." *Scientific Reports* 8 (1): 1–14. https://doi.org/10.1038/s41598-018-22977-4.

Florea, Vlad Alexandru, Andreea Cristina Ionică, Adrian Florea, Răzvan Bogdan Itu, and Mihai Popescu-Stelea. 2022. "Study of the Possibilities of Improving Maintenance of Technological Equipment Subject to Wear." *Processes* 10 (12): 2550. https://doi.org/10.3390/PR10122550.

Gabriel, Barbara, Guy Bodenmann, and Steven R. H. Beach. 2016. "Gender Differences in Observed and Perceived Stress and Coping in Couples with a Depressed Partner." *Open Journal of Depression* 5 (2): 7–20. https://doi.org/10.4236/OJD.2016.52002.

Garcia-Bernabeu, A., J. V. Salcedo, A. Hilario, D. Pla-Santamaria, and Juan M. Herrero. 2019. "Computing the Mean-Variance-Sustainability Nondominated Surface by Ev-Moga." *Complexity* 2019. https://doi.org/10.1155/2019/6095712.

Gopi, Vivek, and P. G. Saleeshya. 2022. "An Integrated Sustainable Lean Approach for the SMEs in India: A Multi-Level Conceptual Frame Work." *Materials Today: Proceedings* 59 (January): 1615–20. https://doi.org/10.1016/J.MATPR.2022.03.155.

Hahn, Gerald J. 1984. "Experimental Design in the Complex World." *Technometrics* 26 (1): 19–31. https://doi.org/10.1080/00401706.1984.10487918.

Haughey, Duncan. 2023. "RACI Matrix." https://www.projectsmart.co.uk/tools/raci-matrix.php.

Hernández-Soto, Roberto, Mónica Gutiérrez-Ortega, and Avi Bartolomé Rubia. 2021. "Revisión Sistemática Exploratoria Sobre El Intercambio de Conocimiento En Las Comunidades de Práctica Virtuales." *Investigação Qualitativa Em Ciências Sociais: Avanços E Desafios // Investigación Cualitativa En Ciencias Sociales: Avances Y D* 9 (July): 239–48. https://doi.org/10.36367/NTQR.9.2021.239–248.

Jankovic, Aleksandar, Gaurav Chaudhary, and Francesco Goia. 2021. "Designing the Design of Experiments (DOE)—An Investigation on the Influence of Different Factorial Designs on the Characterization of Complex Systems." *Energy and Buildings* 250 (November): 111298. https://doi.org/10.1016/J.ENBUILD.2021.111298.

Jaqin, C., A. Rozak, and H. Hardi Purba. 2020. "Case Study in Increasing Overall Equipment Effectiveness on Progressive Press Machine Using Plan-Do-Check-Act Cycle." *International Journal of Engineering* 33 (11): 2245–51. https://doi.org/10.5829/IJE.2020.33.11B.16.

Knop, Krzysztof. 2021. "Management of Packaging Labeling Technology in the Context of Improving the Final Product Quality and Work Safety." *System Safety: Human—Technical Facility—Environment* 3 (1): 116–28. https://doi.org/10.2478/CZOTO-2021-0013.

Knop, Krzysztof, and Robert Ziora. 2022. "Statistical Analysis and Prediction of the Product Complaints." *System Safety: Human—Technical Facility—Environment* 4 (1): 99–115. https://doi.org/10.2478/CZOTO-2022-00011.

Kreager, B. Z., N. D. LaDue, T. F. Shipley, R. D. Powell, and B. A. Hampton. 2022. "Spatial Skill Predicts Success on Sequence Stratigraphic Interpretation." *Geosphere* 18 (2): 750–61. https://doi.org/10.1130/GES02428.1.

Kumar, S., and B. Adaveesh. 2017. "Application of 8D Methodology for the Root Cause Analysis and Reduction of Valve Spring Rejection in a Valve Spring Manufacturing Company: A Case Study." *Indian Journal of Science and Technology* 10 (11): 1–11. https://doi.org/10.17485/ijst/2017/v10i11/106137.

Leapsome. 2024. "Team Celebrations at Work: A Guide for Managers." https://www.leapsome.com/blog/how-to-celebrate-success-at-work.

Li, Xingxing, and Ke Yang. 2020. "Parametric Exploration on the Airfoil Design Space by Numerical Design of Experiment Methodology and Multiple Regression Model." *Proceedings of the Institution of Mechanical Engineers, Part A: Journal of Power and Energy* 234 (1): 3–18. https://doi.org/10.1177/0957650919850426/19850426.

Li, Yang, Cuiyu Wang, Liang Gao, Yiguo Song, and Xinyu Li. 2021. "An Improved Simulated Annealing Algorithm Based on Residual Network for Permutation Flow Shop Scheduling." *Complex and Intelligent Systems* 7 (3): 1173–83. https://doi.org/10.1007/S40747-020-00205-9/TABLES/5.

Lipińska, Milena, Marzena Tomaszewska, and Danuta Kołozyn-Krajewska. 2019. "Identifying Factors Associated with Food Losses during Transportation: Potentials for Social Purposes." *Sustainability* 11 (7): 2046. https://doi.org/10.3390/SU11072046.

Luca, L., and T. O. Luca. 2019. "Ishikawa Diagram Applied to Identify Causes Which Determines Bearings Defects from Car Wheels." In *IOP Conference Series: Materials Science and Engineering*, edited by Eduard Laurenţiu Niţu, Laurenţiu Slatineanu, Monica Daniela Iordache, Adriana Gabriela Plăiaşu, Oana Dodun, Gheorghe Nagîţ, Nicolae Doru Stănescu, et al. Vol. 564, 012093. Pitesti, Romania: IOP Publishing. https://doi.org/10.1088/1757-899X/564/1/012093.

LucidChart. 2023. "Qué Es Un Diagrama de Flujo." *Símbolos de Diagramas de Flujo.* https://www.lucidchart.com/pages/es/que-es-un-diagrama-de-flujo.

Martínez-Martínez, Tere Itzell. 2023. "Simulación de Eventos Discretos Desde La Ingeniería Industrial." *Pädi Boletín Científico De Ciencias Básicas E Ingenierías Del ICBI* 10 (20): 110–21. https://doi.org/10.29057/icbi.v10i20.10207.

Mathiyazhagan, K., A. Gnanavelbabu, Naveen Kumar.N, and Vernika Agarwal. 2022. "A Framework for Implementing Sustainable Lean Manufacturing in the Electrical and Electronics Component Manufacturing Industry: An Emerging Economies Country Perspective." *Journal of Cleaner Production* 334 (February): 130169. https://doi.org/10.1016/J.JCLEPRO.2021.130169.

Mitreva, Elizabeta, Nako Taskov, and Hristijan Gjorshevski. 2015. "Methodology for Design and Implementation of the TQM (Total Quality Management) System in Automotive Industry Companies in Macedonia." In *Regionalna Naucno Strucna Konferencija—ERAZ 2015—Odrzivi Ekonomski Razvoj-Savremeni i Multidisciplinarni Pristup*, 1–9. Belgrade, Serbia: UGD Academic Repository. http://eprints.ugd.edu.mk/14648/.

Mitreva, Elizabeta, Liljana Zdrvkovska, Nako Taskov, Dejan Metodijeski, and Hristijan Gjorshevski. 2016. "Quality Management and Practises in Automotive Parts Production." In *XXIV International Scientific-Technical Conferencxxiv International Scientific and Technical Conference on Transport, Road-Building, Agricultural, Hoisting & Hauling and Military Technics and Technologies*, 4–7. Sofia, Bulgaria. http://eprints.ugd.edu.mk/16861/.

Montgomery, Douglas C. 2001. "Experimental Design for Product and Process Design and Development." *Journal of the Royal Statistical Society: Series D (The Statistician)* 48 (2): 159–77. https://doi.org/10.1111/1467-9884.00179.

Morell-Santandreu, Oscar, Cristina Santandreu-Mascarell, and Julio J. Garcia-Sabater. 2021. "A Model for the Implementation of Lean Improvements in Healthcare Environments as Applied in a Primary Care Center." *International Journal of Environmental Research and Public Health* 18 (6): 2876. https://doi.org/10.3390/IJERPH18062876.

Muncut, E. S., L. I. Culda, G. M. Erdodi, and G. Sima. 2019. "8D Complaint Solving Method in an Automotive Component Processing Company." In *IOP Conference Series: Materials Science and Engineering*, edited by Gavril Grebenişan and Alexandru-Viorel Pele. Vol. 568, 012020. Oradea: IOP Publishing. https://doi.org/10.1088/1757–899X/568/1/012020.

Nagamine, Yoshitaka, Ricardo Pong-Wong, Peter M. Visscher, and Chris S. Haley. 2009. "Detection of Multiple Quantitative Trait Loci and Their Pleiotropic Effects in Outbred Pig Populations." *Genetics Selection Evolution* 41 (1): 1–11. https://doi.org/10.1186/1297-9686-41-44/TABLES/8.

Nicolae, Viorel, Laurenţiu Mihai Ionescu, Nadia Belu, and Ştirbu Luminiţa Elena. 2017. "Improvement of the 8D Analysis Through a System Based on the 'Internet of Things' Concept Applied in Automotive Industry." In *CONAT 2016 International Congress of Automotive and Transport Engineering*, edited by A. Chiru and N. Ispas, 635–42. Cham: Springer International Publishing. https://doi.org/10.1007/978-3-319-45447-4_70.

NSB, Akhil, Vimal Kumar, Tanmoy De, and Suriya Kalangrit. 2022. "Quality Function Deployment Analysis of Smartphones." *Teknomekanik* 5 (2): 72–79. https://doi.org/10.24036/TEKNOMEKANIK.V5I2.14372.

Omnex. 2024. "8D Problem Solving | Benefits, Process & Tools." *What Is Problem Solving?* https://www.omnex.com/problem-solving.

Ortiz-Hernandez, Carlos, Alexander Troncoso-Palacio, Dylan Acosta-Toscano, Robert Begambre-Meza, and Blass Troncoso-Mendoza. 2019. "Utilización de Herramientas de Calidad Para La Mejora En Los Procesos de Extrusión de Plásticos." *Boletín de Innovación, Logística y Operaciones* 1 (1): 1–7. https://doi.org/10.17981/BILO.01.01.2019.01.

Øvretveit, John. 1995. "Team Decision-Making." *Journal of Interprofessional Care* 9 (1): 41–51. https://doi.org/10.3109/13561829509010448.

Pacana, Andrzej, and Karolina Czerwińska. 2021. "Model of Diagnosing and Searching for Incompatibilities in Aluminium Castings." *Materials* 14 (21): 6497. https://doi.org/10.3390/MA14216497.

Pacheco-Pacheco, Víctor Hugo. 2018. "Implementación de Un Modelo de Mejora Para Optimizar El Tiempo de Entrega de Los Productos 'Alto De Basta' y 'Alto De Camisa' En El Proceso Confección, Arreglo y Modificación, Mediante La Aplicación de Las 8 Disciplinas. Caso: Symp—Sastrería." Universidad Católica del Ecuador. http://repositorio.puce.edu.ec/handle/22000/14869.

Pazhayattil, Ajay Babu, Naheed Sayeed-Desta, Emilija Fredro-Kumbaradzi, and Jordan Collins. 2018. "Life Cycle Approach to Process Validation." In *Solid Oral Dose Process Validation*. 1st ed., 1–7. Cham, Switzerland: Springer. https://doi.org/10.1007/978-3-030-02472-7_1.

PMOinformatica.com. 2015. "Flujograma de Procesos y Gerencia de Proyectos." https://www.pmoinformatica.com/2015/07/flujograma-de-procesos-gerencia.html.

Pro Academic Writers. 2024. "Types of Research Design—Strong Fundamentals." *Get Acquainted with Types of Research Design—Succeed in Research Paper Writing.* https://pro-academic-writers.com/blog/types-of-research-design.

Pulido-Rojano, Alexander D., Alex Ruiz-Lázaro, and Luis Eduardo Ortiz-Ospino. 2020. "Mejora de Procesos de Producción a Través de La Gestión de Riesgos y Herramientas Estadísticas." *Ingeniare. Revista Chilena de Ingeniería* 28 (1): 56–67. https://doi.org/10.4067/S0718-33052020000100056.

QIT Consulting. 2024. "QIT Solutions." https://www.qitconsulting.com/solutions/.

Realyvásquez-Vargas, Arturo, Karina Cecilia Arredondo-Soto, Jorge Luis García-Alcaraz, and Emilio Jiménez-Macías. 2020. "Improving a Manufacturing Process Using the 8Ds Method. A Case Study in a Manufacturing Company." *Applied Sciences* 10 (7): 2433–58. https://doi.org/10.3390/app10072433.

Rodrigues dos Santos, Anderson, Flávio de São Pedro Filho, Fabricio Moraes de Almeida, Marcelo José Peres Gomes da Silva, and Tiyao Sui-Qui. 2018. "Innovating Management Control by Dynamic Analysis of Pareto in a Hotel Business." *International Journal of Advanced Engineering Research and Science* 5 (4): 197–206. https://doi.org/10.22161/IJAERS.5.4.29.

Rodriguez, Jackelyn. 2016. "Step 9: Verification of Effectiveness." In *CAPA in the Pharmaceutical and Biotech Industries*, edited by Jackelyn Rodriguez, 159–70. Cambridge: Woodhead Publishing. https://doi.org/10.1016/B978-1-907568-58-9.00012-6.

Rodríguez, Ligia, and Shirley Rodríguez. 2004. "Methodological Process to Plan and Implement an Educational Campaign to Increase Bean Consumption." *Agronomía Mesoamericana* 15 (3): 327–41. https://doi.org/10.15517/am.v15i3.11885.

Rodríguez-Palomino, Paulino. 2021. "Inteligencia Artificial Para La Administración de Los Reglamentos Aeronáuticos De Colombia (RAC)." *Revista Estrategia Organizacional* 10 (1): 1–27. https://doi.org/10.22490/25392786.4551.

Roque, Raymundo, and Elkhy Berenice. 2017. *Reingeniería de Procesos En La Fabricación de Unidades Odontológicas Aplicando La Metodología de Las Ocho Disciplinas.* Universidad Tecnológica de Perú.

Sadiq, Saba, Muhammad Saad Amjad, Muhammad Zeeshan Rafique, Shafqat Hussain, Uzma Yasmeen, and Mohammad Aamir Khan. 2021. "An Integrated Framework for Lean Manufacturing in Relation with Blue Ocean Manufacturing—A Case Study." *Journal of Cleaner Production* 279 (January): 123790. https://doi.org/10.1016/J.JCLEPRO.2020.123790.

Sadiq, Shazia, Maria Orlowska, Wasim Sadiq, and Cameron Foulger. 2004. "Data Flow and Validation in Workflow Modelling." In *Proceedings of the 15th Australasian Database Conference*, edited by Klaus-Dieter Schewe and Hugh Williams, 207–14. Dunedin, New Zealand: Australian Computer Society. https://dl.acm.org/doi/abs/10.5555/1012294.1012317.

Sahu, Neelesh Kumar, and Atul Andhare. 2018. "Design of Experiments Applied to Industrial Process." In *Statistical Approaches with Emphasis on Design of Experiments Applied to Chemical Processes*, edited by Valter Silva, 1–180. IntechOpen. https://doi.org/10.5772/INTECHOPEN.73558.

Sánchez Esquerra, Maricarmen, and José Lozano Taylor. 2001. *Aplicación de Las Siete Herramientas Básicas de La Calidad Al Proceso de Costura En La Empresa Lear Corporation Hermosillo*. Hermosillo: Universidad de Sonora. http://repositorio-institucional.uson.mx/handle/20.500.12984/3269.

Santosa, Iqbal, and Rahmat Mulyana. 2023. "The IT Services Management Architecture Design for Large and Medium-Sized Companies Based on ITIL 4 and TOGAF Framework." *JOIV: International Journal on Informatics Visualization* 7 (1): 30–36. https://doi.org/10.30630/JOIV.7.1.1590.

Sauer, W., M. Opperman, G. Weigert, S. Werner, H. Wohlrabe, K.-J. Wolter, and T. Zerna. 2006. "Design of Experiments and Regression Analyses." In *Electronics Process Technology*, edited by W. Sauer, M. Opperman, S. Werner, H. Wohlrabe, K. LastName, and T. Zerna, 349–87. London: Springer. https://doi.org/10.1007/1-84628-354-X_7.

Schmitt, Charles P., and Margaret Burchinal. 2011. "Data Management Practices for Collaborative Research." *Frontiers in Psychiatry* 2 (July): 1–8. https://doi.org/10.3389/FPSYT.2011.00047/BIBTEX.

Sitko, Jacek, Katarzyna Midor, and T. N. Ivanova. 2021. "Analysis the Causes of Quality Defects in a Car Rack." *E3S Web of Conferences* 285 (07036): 1–10. https://doi.org/10.1051/e3sconf/202128507036.

Six Sigma. 2017. "What Is Ishikawa (Fishbone Diagram)?" https://www.6sigma.us/etc/what-is-ishikawa-fishbone-diagram/.

Six Sigma. 2024. "Root Cause Analysis (RCA) for Beginners: What Is Root Cause Analysis (RCA)?" https://www.6sigma.us/etc/root-cause-analysis-for-beginners/.

Škůrková, Katarína Lestyánszka. 2017. "Implementation of a System Quality Tool to Reduce The Costs of Scrap Loss in Industrial Enterprise." *Zeszyty Naukowe Quality. Production. Improvement* 1 (6): 93–111. http://yadda.icm.edu.pl/yadda/element/bwmeta1.element.baztech-fa47b9da-833a-40ff-b3ee-42b79ae34e86.

Stepanek, Lubomir, Filip Habarta, Ivana Mala, and Lubos Marek. 2022. "A Short Note on Post-Hoc Testing Using Random Forests Algorithm: Principles, Asymptotic Time Complexity Analysis, and Beyond." In *Proceedings of the 17th Conference on Computer Science and Intelligence Systems*, edited by M. Ganzha, L. Maciaszek, M. Paprzycki, and D. Ślęzak, 489–97. Sofia, Bulgaria: FedCSIS. https://doi.org/10.15439/2022F265.

Talancon, Cecilia Michelle, Josué López-Leyva, Dalia Chávez-García, Miguel Ponce-Camacho, and Ariana Talamantes-Álvarez. 2020. "A Novel Space Systems Management Methodology Based on Shortcomings and Strengths of Conventional System Engineering Tools Used in a Design Thinking Framework." In *Lean Manufacturing and Six Sigma—Behind the Mask*, edited by Fausto Pedro García-Márquez, Isaac Segovia-Ramirez, Tamás Bányai, and Péter Tamás, 1–19. IntechOpen. https://doi.org/10.5772/INTECHOPEN.83430.

Tanco, Martin, Elisabeth Viles, Laura Ilzarbe, and Maria Jesus Alvarez. 2009. "Implementation of Design of Experiments Projects in Industry." *Applied Stochastic Models in Business and Industry* 25 (4): 478–505. https://doi.org/10.1002/ASMB.779.

Tian, Jie, Zhibo Han, Heye Reemt Bogena, Johan Alexander Huisman, Carsten Montzka, Baoqing Zhang, and Chansheng He. 2020. "Estimation of Subsurface Soil Moisture from Surface Soil Moisture in Cold Mountainous Areas." *Hydrology and Earth System Sciences* 24 (9): 4659–74. https://doi.org/10.5194/HESS-24-4659-2020.

Titu, Mihail Aurel. 2016. "Management of Customers' Complaints within SC COMPA SA Sibiu. Case Study for a Topical Complaint." *Journal of Electrical Engineering, Electronics, Control and Computer Science* 2 (4): 1–4. https://jeeeccs.net/index.php/journal/article/view/42.

Torres Sosa, Jesús Antonio, Luis Alberto Rodríguez Picón, Luis Carlos Méndez González, and Iván Juan Carlos Pérez Olguín. 2020. "Diseño de Experimentos Para Optimizar Resistencia e Índices de Capacidad de Un Fusible." *Cultura Científica y Tecnológica* 17 (2): 1–9. https://doi.org/10.20983/culcyt.2.2.4.

Tripathi, Varun, Somnath Chattopadhyaya, Alok K. Mukhopadhyay, Shubham Sharma, Jujhar Singh, Danil Yurievich Pimenov, and Khaled Giasin. 2021. "An Innovative Agile Model of Smart Lean–Green Approach for Sustainability Enhancement in Industry 4.0." *Journal of Open Innovation: Technology, Market, and Complexity* 7 (4): 215. https://doi.org/10.3390/JOITMC7040215.

Valier, Alison R. Snyder. 2020. "Looking to Improve Your Practice? Consider the Science of Quality Improvement to Get Started." *Journal of Athletic Training* 55 (11): 1137–41. https://doi.org/10.4085/1062-6050-0342.19.

Vallejo-Castillo, Vladimir, Aida Rodríguez-Stouvenel, Ronny Martínez, and Claudia Bernal. 2020. "Development of Alginate-Pectin Microcapsules by the Extrusion for Encapsulation and Controlled Release of Polyphenols from Papaya (Carica Papaya L.)." *Journal of Food Biochemistry* 44 (9): e13331. https://doi.org/10.1111/JFBC.13331.

Vargas, Diego Leandro. 2017. "Resolução de Problemas Utilizando a Metodologia 8D: Estudo de Caso de Uma Indústria Do Setor Sucroalcooleiro." In *Simpósio de Engenharia de Produção de Sergipe Anais Do IX SIMPROD*, 464–77. São Cristóvão: Departamento de Engenharia de Produção—Universidade Federal de Sergipe. https://ri.ufs.br/jspui/handle/riufs/7672.

Vencúrik, Tomas, Damir Knjaz, Tomislav Rupčić, Goran Sporiš, and Feng Li. 2021. "Kinematic Analysis of 2-Point and 3-Point Jump Shot of Elite Young Male and Female Basketball Players." *International Journal of Environmental Research and Public Healt* 18 (3): 934. https://doi.org/10.3390/IJERPH18030934.

Vidal, Cristian L., Dannia D. Hernández, Cristian A. Pereira, and María C. Del Río. 2012. "Aplicación de La Modelación Orientada a Aspectos." *Información Tecnológica* 23 (1): 3–12. https://doi.org/10.4067/S0718-07642012000100002.

Wang, Xiuzhen, Zhaoxia Xu, Lianjie Jiang, Chuanzhi Sun, Jun Yan, and Li Gao. 2022. "Importance Analysis of Structural Seismic Demand Based on Support Vector Machine." *Mathematical Problems in Engineering* 2022: 1–14. https://doi.org/10.1155/2022/1283105.

Wichawong, Parinya, and Prabhas Chongstitvatana. 2017. "Knowledge Management System for Failure Analysis in Hard Disk Using Case-Based Reasoning." In *2017 18th IEEE/ACIS International Conference on Software Engineering, Artificial Intelligence, Networking and Parallel/Distributed Computing (SNPD)*, 1–6. Kanazawa, Japan: IEEE. https://doi.org/10.1109/SNPD.2017.8332383.

Wilkinson, Leland. 2006. "Revising the Pareto Chart." *The American Statistician* 60 (4): 332–34. https://doi.org/10.1198/000313006X152243.

Worsfold, Paul J., Eric P. Achterberg, Antony J. Birchill, Robert Clough, Ivo Leito, Maeve C. Lohan, Angela Milne, and Simon J. Ussher. 2019. "Estimating Uncertainties in Oceanographic Trace Element Measurements." *Frontiers in Marine Science* 6 (January): 387497. https://doi.org/10.3389/FMARS.2018.00515/BIBTEX.

Yirga, Ashenfai A., Sileshi F. Melesse, Henry G. Mwambi, and Dawit G. Ayele. 2020. "Modelling CD4 Counts before and after HAART for HIV Infected Patients in KwaZulu-Natal South Africa." *African Health Sciences* 20 (4): 1546–61. https://doi.org/10.4314/AHS.V20I4.7.

Zasadzień, Michał. 2016. "Wykorzystanie Metody 8D Do Doskonalenia Procesu Remontowego Wagonów Kolejowych." *Systemy Wspomagania w Inżynierii Produkcjiz* 2 (14): 392–99. https://www.infona.pl/resource/bwmeta1.element.baztech-3be87d67-399f-477c-bc50-6f3e623b56e2.

Zasadzień, Michał. 2017. "Six Sigma Methodology as a Road to Intelligent Maintenance." *Production Engineering Archives* 15: 45–48. http://yadda.icm.edu.pl/yadda/element/bwmeta1.element.baztech-a5a5d776-391b-46bc-9712-80cc2b2a608a.

Zhou, Fuli, Xu Wang, Tshepo Mpshe, Yi Zhang, and Yuhang Yang. 2016. "Quality Improvement Procedure (QIP) Based on 8D and Six Sigma Pilot Programs in Automotive Industry." In *Proceedings of the First International Conference Economic and Business Management*, edited by W.C. Du, 275–81. Qingdao, China: Atlantis Press. https://doi.org/10.2991/FEBM-16.2016.43.

Zhou, Ning, Qian Ren, and Jianxin Zhou. 2020. "Identification of Critical-to-Quality Characteristics Based on Improved XGBoost." In *DSIT 2020: Proceedings of the 3rd International Conference on Data Science and Information Technology*, edited by Yuan Yan Tang, Tok Wang Ling, and Hee Yong Youn, 205–09. Xiamen, China: Association for Computing Machinery. https://doi.org/10.1145/3414274.3414508.

3 Case Study 1. Validation of the Seal Ring Annealing Process

3.1 INTRODUCTION

Failure to validate manufacturing processes and measuring equipment can have a negative impact on a company. Among these consequences are, according to Ahir et al. (2014), low-quality products, increased quality costs, low process efficiency, and malfunctioning measurement equipment. In the case of low-quality products, these arise because, in the absence of a validated production process, there is no control over the process; therefore, there is no guarantee that the process consistently generates quality products. However, an unvalidated manufacturing process will lead to rejects, rework, retesting, and more inspections. All of this represents the company's costs. Low efficiency goes hand in hand with the first two consequences, as the efficiency of a process results in products that meet quality requirements at the lowest cost. Finally, not having validated equipment, that is, calibrated, can result in erroneous measurements of critical variables in product quality. Other consequences include lower productivity and longer cycle time (Tripathi et al. 2021).

Different cases in the literature exemplify these consequences in non-validated production processes. For example, Lefeber et al. (2004) discussed the case of a semiconductor production process. In this process, owing to the large

DOI: 10.1201/9781003650966-3

number of process steps, the reentrant nature of the process flow, and the use of advanced technologies, wafer fabrication, and production times are relatively long, exceeding two months. Therefore, problems arise with production and on-time delivery, as it is still being determined when and how many semiconductors can be fabricated. To solve this problem, the authors developed and validated a new model.

In contrast, Psomas et al. (2013) pointed out that companies must continuously improve their quality management systems (QMS). To achieve this, they must implement the ISO 9001 standard, which aims to provide an effective QMS that reflects a company's quality practices in producing goods and services that conform to requirements. However, there needs to be an instrument to measure the effectiveness of this standard, especially in the food sector. Therefore, Psomas et al. (2013) developed and validated an instrument to measure the effectiveness of ISO 9001 based on its components, namely, ISO 9001 objectives, in the food manufacturing sector. Thus, based on the current degree of achievement of ISO 9001 objectives, a food company can further improve the quality of its processes and products, move closer to business excellence, and maximize its respective profits.

Similarly, Shabi et al. (2017) developed a project to validate the process of software design, hardware development, and avionics upgrade of an existing system in the Israeli Aircraft Industry (IAI), as well as to add some additional operational capabilities. The IAI deployed a new system at the customer's site to minimize the operational system shutdown duration. Therefore, it is essential to perform a complete validation of the process before deployment to allow customers to evaluate it. This evaluation included validation of customer requirements and verification of engineering requirements. Shabi et al. (2017) validated the process by developing a model to optimize quality based on cost, risk, and process time, ensuring that the customers receive high-quality products.

This background clarifies that the validation of manufacturing processes plays an essential role in product quality, production times, and costs. Next, a case study on the validation problem of a manufacturing company's sealing ring annealing process was presented. Subsequently, the objectives are presented, and the problem is solved by applying the 8Ds methodology, along with auxiliary tools and results.

3.2 CASE STUDY

This study was conducted at a Japanese company located in Tijuana, Mexico. This company is dedicated to manufacturing electronic and ceramic components and has expanded its presence worldwide through subsidiaries in the America, Europe, and Asia. It is a global company that manufactures and sells a wide variety of products, including printers and multifunctional equipment, ceramic components, electronic devices, automotive components, semiconductors, industrial tools, LCD products, printing equipment components, optical components, solar electric systems, kitchen products, and office products.

Plants located in Tijuana mainly manufacture high-quality ceramic components for various industrial applications, including the electronics, automotive, aerospace, medical, and renewable energy industries. The plant has different departments: maintenance, human resources, quality, and transistor outline (TO). In the specific case of the TO department, it is one of the leading departments responsible for manufacturing SMD packages in the company. The department focuses on the production of high-quality electronic components. To do so, it carries out a rigorous process that involves different work areas, including production line supervision, quality management, and research and development of new technologies.

The TO organization chart is divided into different hierarchical levels, from managers and line managers to operators and technicians who work together to ensure quality and efficiency in the manufacturing process of electronic components. In this team, the engineer figure is fundamental because it is responsible for the process validation and optimization.

The functions of the members of the OT department are as follows:

- Production manager: He oversees and coordinates the company's manufacturing and production operations, ensuring that delivery schedules are met, quality standards are met, and resources are used optimally and cost-effectively.
- Assistant Production Manager: He is responsible for supporting and assisting the production manager in supervising and coordinating the company's manufacturing and production operations. His purpose is to ensure that production processes are efficient, effective, and profitable and that delivery deadlines and quality standards are met.
- Production Supervisor: He is responsible for reviewing and coordinating the daily activities of the company's employees. His role is to ensure that the established objectives and goals are met and that quality and safety standards are maintained in the workplace. He aims to motivate and lead his work team to achieve maximum efficiency and productivity.
- Line Manager: He is responsible for coordinating and supervising activities on a specific production line in the company. His role is to ensure that production processes are efficient, quality standards

are met, and established production targets are achieved. He aims to lead and motivate production line workers to improve efficiency and productivity in his work area.

- Engineers: They are responsible for designing, developing, and improving the company's production processes. Their function is to analyze current processes, identify problems, and propose solutions to improve efficiency and quality. They aim to optimize production processes to maximize profitability and satisfy customer needs.
- Engineering Inspector: He is responsible for verifying the quality of finished products before shipment to the customer, ensuring that they meet established quality standards. Its purpose is to guarantee customer satisfaction and the company's reputation.
- Production Technician: He oversees rework and lot closure and ensures that products not meeting quality standards are repaired or disposed of.
- Practitioner: His/her purpose is to learn and gain experience in the production area while performing assigned tasks.
- Production Personnel: Responsible for executing assigned tasks in the production process to achieve established objectives.
- Engineering Assistant: His/her purpose is to provide administrative and technical support to the production team.

The Tijuana plant has several areas dedicated to specific tasks during the manufacturing of electronic components. These areas include assembly, soldering, inspection, shelving, engineering, quality, and offices. The functions of each area in the Tijuana plant are as follows:

- Assembly: Its function is to precisely join product components, ensuring that customer specifications are met and the quality of the final product is assured.
- Soldier: Its primary function is to join the components of the products through welding processes. Its purpose is to ensure the final product's quality and comply with the standards established by the customers.
- Inspection: Its function is to guarantee the quality of products through tests and controls. Its purpose is to ensure that products meet established quality standards and exceed customer expectations.
- Shelving: Its primary function is to store and organize finished products. Its purpose is to ensure that products are available for shipment and maintained in optimal conditions until delivery to the customer or to the various processes that utilize subcomponents.

- Engineering: It is responsible for developing and improving manufacturing processes. Its purpose is to optimize processes to improve the efficiency and quality of production and reduce costs. Likewise, they are responsible for providing solutions and answers to the different problems that arise in the plant daily and delivering the results of processes for the follow-up of their performance.
- Quality: It is responsible for ensuring that products meet established quality standards. Its purpose is to maintain customer satisfaction and the company's reputation, as well as to continuously improve production processes.
- Administrative Offices: They manage the company's human, financial, and material resources. Their purpose is to ensure the company's efficient operation and the fulfillment of its established objectives, enabling the organization to grow sustainably and profitably.

Each area is crucial for producing high-quality products that characterize the company. The TO department in the soldering area validated the annealing process for the sealing rings (in three categories: A, B, and C) of the SMD packages. Currently, this area only performs component joining by soldering and sintering heat treatment for various numbers of parts. Previously, this process was performed at a plant located in San Diego, California, and after annealing, the material was returned to the plant in Tijuana.

The specialized machinery used to execute the annealing process is a high-performance furnace that rapidly achieves high temperatures and maintains the heating curve for extended periods. As part of the process validation, it is required that it be capable of being executed by production, as well as the optimization of the task to supply in time and form the quantities of material required by the customer, in this case, the TO department itself. Activities are performed throughout the three shifts to cover washing functions, weld inspection, package rework, and loading and unloading material in ovens.

Figure 3.1 shows the process flow diagram for the annealing process of the sealing rings for SMD packages. This process is divided into six main steps: degreasing, annealing, inspection, plating, sintering, and storage. The material handling process for the seal ring consists of preparing the material for cleaning. This material is sent to the San Diego plant for heat treatment and then returned to the Tijuana plant, where it undergoes additional inspection, plating, measurement, sintering, and storage processes.

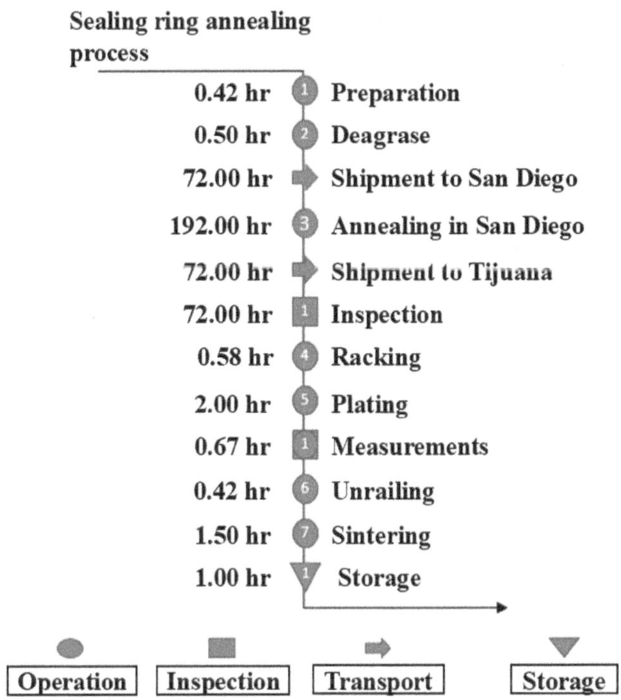

FIGURE 3.1 Sealing ring annealing process flow diagram

3.2.1 Problem Statement

Currently, the ring-annealing process is not mass-produced by the Tijuana plant. Therefore, we aim to implement it to optimize its time and efficiency and offer electronic components of the highest quality and reliability. During the soldering process, as well as during the assembly of SMD packages, cracks, damage, deformities, and misalignment can occur in the sealing rings, which affect their durability and resistance. In addition, damage to the surface of the rings, such as scratches and dents, compromising their tightness and performance, has been identified. Ring deformation during the manufacturing process can lead to misalignment and installation difficulties. The variability in the dimensions of the rings can affect their fit and tightness. These problems represent a risk to the quality and reliability of the packages, so it is necessary to find solutions to improve the manufacturing process of the sealing rings, as these are critical components to ensure the functionality and reliability of electronic devices. Over 11 months, from January to November 2022, 185 sealing ring lots were processed, corresponding to a total of 220,439 pieces, with an average value of US$265 thousand, which were prepared for SMD assembly.

The defects present in the rings correspond mostly to missing materials (i.e., missing parts) and damaged parts. Engineering tests and missing material defects were not considered in the development of this validation, given the nature of the rejection in the process that was used for the failed validation of a previous annealing process. Therefore, as shown in the Pareto diagram in Figure 3.2 (A), 5,185 defects were recorded in 2022, of which 80% corresponded to exposed metal, damaged parts, foreign material, and accumulations. The exposed metal and damaged parts are defects that are directly related to the sealing ring and their ability to adhere or malleability. On the other hand, the defects of each type of SMD recorded in 2023 are shown in Figure 3.2 (B), Figure 3.2 (C), and Figure 3.2 (D). It can be seen that of the 100% (60353) of the processed parts that reached assembly as SMD, the package with the highest level of rejection is SMD A, with 10.92% (3336 parts) of defects. For SMD B and SMD C, 715 and 1,133 defects were obtained, respectively. This indicates that the process does not produce high-quality parts and must be validated.

In April 2023, 50% of TO materials were processed within the last three days of the production plan. This represents a bottleneck that requires overtime, parallel validation processes, and the implementation of containment actions to solve the problems that arise. The SMD demand corresponds to 10–12 batches per month, with 1,200 parts per batch, equivalent to 12,000–14,400 parts per month. Among the part numbers processed in April 2023, 18% correspond to SMD lots, representing a total of 11,829 pieces. The SMDs are classified into three parts (A, B, and C), each with a preferred sealing ring type. For the quantities of processed SMD, none of the three parts met 100% of the demand, and the average rejection rate was 7.72%. This indicates that the annealing process must be validated.

The SMD demand corresponds, on average, to 754 pieces per day. Given the increase in SMD production required to meet total demand, it is expected to rise from 12 to 19 batches per month, exceeding the initial number of sealing rings processed. In other words, it will require 754–1,140 pieces of SMD per day. In addition, to maintain the margin of the sealing ring pieces in the process, corresponding to a 16.84% slack, the production of 1,140 SMD of sealing rings is required, corresponding to 1,332 pieces per day.

Table 3.1 shows the values that each activity adds to the process, as shown in Figure 3.1. These activities are classified as those that generate value for the customer (VAC), those that generate value for the company (VAE), and those that do not. The latter are further classified into preparation (P), delay (D), movement (M), control (C), and archiving (A).

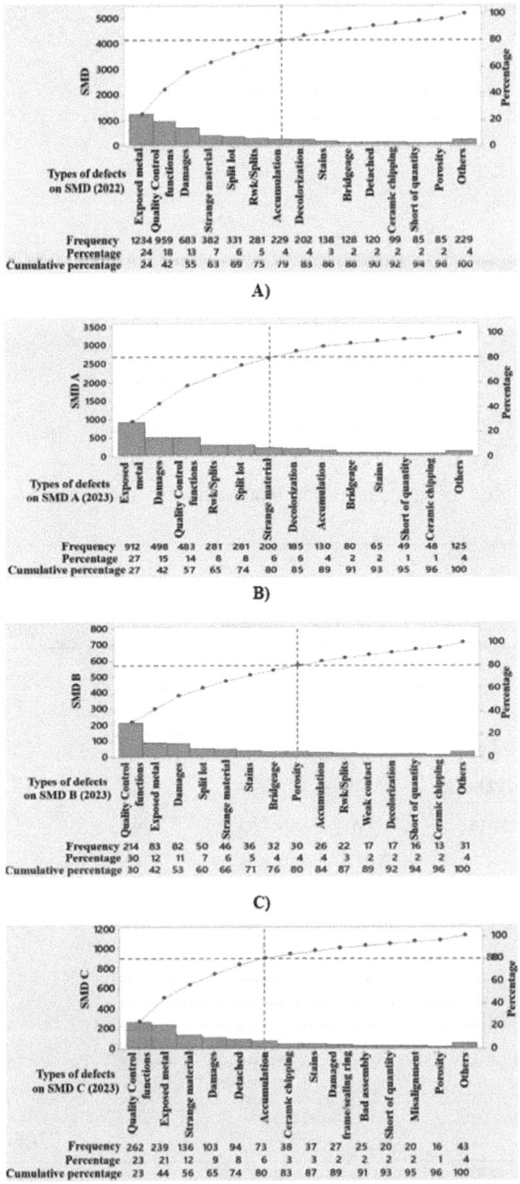

FIGURE 3.2 Pareto diagrams of defects of sealing rings: (A) Defects recorded in 2022, (B) Defects of SMD A in 2023, (C), Defects of SMD B in 2023, and (D) Defects of SMD C in 2023.

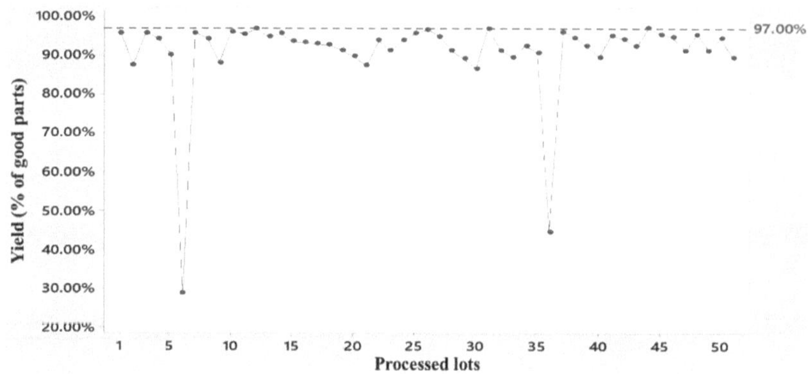

FIGURE 3.3 SMD batch performance

TABLE 3.1 Matrix of classification, value-added, and time of the annealing process activities

DESCRIPTION	*RANKING*	*ADD VALUE*	*IT IS NECESSARY TO*	*TIME (MIN)*	*%*
Preparation	P	No	Yes	25	0.121%
Degreasing	C	No	No	30	0.145%
Shipment to San Diego	M	No	No	4320	20.925%
Annealing in San Diego	VAE	Yes	No	11520	55.800%
Shipment to Tijuana	M	No	No	4320	20.925%
Inspection	C	No	No	60	0.291%
Enraque	M	No	Yes	35	0.170%
Platinum	VAC	Yes	Yes	120	0.581%
Measurements	C	No	No	40	0.194%
Unraveling	M	No	Yes	25	0.121%
Sintered	VAE	Yes	Yes	90	0.436%
Storage	A	No	Yes	60	0.291%
Total				20645	100%

A total of 20,645 minutes represents a delivery problem with the following assembly process in the SMD chain, as most of the materials were processed in the last three days of the month. Of these, 56.24% represent value added to the company, suggesting that most of the activities in this process add value.

On the other hand, movements represent 42.14% of the total time, suggesting the need to improve the efficiency of the process.

3.2.2 Objectives

The general objective of this project is to validate, using the 8Ds methodology, the annealing process for SMD packages in the TO department. On the other hand, the specific objectives are the following:

- Supply 100% of the sealing ring demand.
- Reduces the preparation process time for ring sealing by at least 82%.
- Increase the batch yield by at least 7% for SMD packages.

3.3 MATERIALS AND METHODS

3.3.1 Materials

Several statistical and industrial engineering tools were used to validate the annealing furnace by applying the 8Ds methodology. These tools include Pareto diagrams, flowcharts, ANOVA, correlation and regression analyses, and Ishikawa diagrams. In addition, various equipment and tools were used to collect and analyze data throughout the process. The purposes of each are described as follows.

- Systems, applications, and products: This tool is used to maintain a detailed record of all activities related to the annealing process. Every movement recorded along the supply chain is logged to keep up-to-date locations, defects, and movements made. This includes creating work orders for equipment maintenance, managing production schedules, and identifying and tracking materials used.
- Scanning electron microscopy (SEM) was used to observe the microstructure of the samples after heat treatment. The samples were examined at a high resolution using SEM to determine the presence of possible defects, such as cracks, porosities, or inclusions. In addition, it was used to measure the grain size and porosity of the sample structure after the heat treatment.

- Keyence microscope: Used to capture high-quality images at magnifications of up to 200×, allowing detailed observation of the characteristics of the annealed parts and verification of their compliance with the required specifications.
- Desktop computer: This was used to record and process the data obtained by the thermocouple, which allowed the temperature inside the oven to be measured. In addition, specialized software integrated into the furnace interprets the data and generates graphs that allow the visualization of the evolution of the temperature inside the furnace during the entire validation process.

3.3.2 Method

The following sections describe how each of the eight disciplines was applied in this project.

3.3.2.1 Stage 1: Establishment of the Teamwork

To create the work team, two tables were developed: one that synthesizes information on the profiles of the team members and another that integrates the dates of the meetings held during the project period. A careful selection of team members was made considering their skills and experience in the project area. Subsequently, the roles and responsibilities of each person were designated. Finally, regular team meetings were established to discuss the project's progress and problems and to maintain constant and effective communication among members.

3.3.2.2 Stage 2: Problem Definition

This phase describes the variables that affect the process and how they imply restrictions. The characteristics and properties to be safeguarded or modified are described in a flow sheet. The problems and deviations that affect these characteristics and objectives, such as the temperature profile, grain size, and ring hardness, are defined. In addition, the impact of the validation on other areas and the considerations to be considered are described. Subsequently, the impact of the problem on the project and the organization as a whole is determined, as well as the level of process and sealing ring demand that sets the standard to be processed.

In addition, the essential elements of the installation and the control variables are at the criterion of the tests, that is, aspects that can be modified to obtain the best result. We recorded and processed data on the temperature profiles (both control and furnace results) and checked the furnace cables and connections. These activities provide a clearer view of the problem and possible solutions that can be implemented to solve it. Necessary adjustments are made by the following standards: NOM-001-SEDE-2012 (DOF—Diario Oficial de la Federación 2012), NOM-002-STPS-2010 (DOF—Diario Oficial de la Federación 2010), NOM-026-STPS-2008 (DOF—Diario Oficial de la Federación 2008), NOM-030-STPS-2009 (DOF—Diario Oficial de la Federación 2009), NOM-057-SEMARNAT-1993 (Mexican Chamber of the Construction Industry 2023), and NMX-E-255-CNCP-2008 (Asociación Mexicana de Constructores de Invernaderos 2013). Please refer to the links in the corresponding references for a more detailed review of the standards.

3.3.2.3 Stage 3: Problem Containment

Using the tower used at the San Diego plant to load the furnace (called Tower A), possible temporary solutions were identified to allow activities to be carried out, that is, a safe method of loading the furnace. In addition, the risks associated with the machine were evaluated, and identified temporary solutions were implemented.

3.3.2.4 Step 4: Root Cause Identification

At this stage, the root cause of the problem was identified. To achieve this, an Ishikawa diagram was used. This tool allows visualization of the different categories of causes that may contribute to the problem, such as people, machines, materials, methods, measurements, and the environment.

3.3.2.5 Step 5: Development of Permanent Corrective Actions

At this stage, a new loading method was designed (Tower B was implemented) based on the constraints identified in the previous stages and the capacity of the operators with respect to the physical force that can be safely exerted. The load should not exceed 35 kg. Subsequently, the quantities of material to be processed for each part number were defined to satisfy the demand requirements.

3.3.2.6 Step 6: Corrective Action Implementation and Validation

At this stage, the effectiveness of the implemented solutions was monitored to ensure that they were working properly and that there was no further propagation of the problem. For this purpose, the weights of Tower A and Tower B are analyzed to determine which of the two provides the maximum efficient load. A Pearson correlation test was performed to measure the level of similarity and relationship between the control and furnace heating curves. For this purpose, Equation (2.1) was applied. The data of the curves were taken as Cartesian points representing the relationship between these two variables over time. To compare the grain size and hardness between the validation batches and the control batch, the ANOVA statistic for multiple range tests was used to determine if the batches presented similarities. The parameters required in ANOVA can be calculated by applying the corresponding parameters in Chapter 2.

The multiple range test was applied to measure the level of similarity between the control and validation batches so that the differences or similarities between them could be appreciated and to define how close or distant they were from each other. For this purpose, Equation (3.1) was applied:

$$\left|\overline{Y_i} - \overline{Y_j}\right| > T_{\frac{\alpha}{2},N-K}\sqrt{CM_{error}\left(\frac{1}{n_1} + \frac{1}{n_j}\right)} \tag{3.1}$$

where $\overline{Y_i}$ is the average Y data of the treatment, $\overline{Y_j}$ is the average Y data of the treatment to be compared, $T_{\frac{\alpha}{2},N-K}$ is the distribution value of the *t-test* statistic, CM_{error} is the mean squares of error, n_1 is the number of sample observations per treatment, and n_j is the number of treatments

The quality control tests performed during validation are described as follows.

- Leak test: This test verifies the ability of the sealing ring to maintain tightness and prevent the leakage of liquids or gases.
- Heater block heating: In this test, the sealing ring was subjected to elevated temperatures to evaluate its thermal resistance and ensure that it did not suffer deformation or damage.
- Seam seal: The quality of the ring seal was evaluated to ensure that it satisfied the required standards.
- Temperature cycling: The sealing ring was subjected to repeated cycles of extreme temperatures to verify its resistance and stability during use.
- Gross leak test and fine leak test: Specific tests are performed to detect any leakage in the seal ring at different pressure levels.
- Dye penetrant: A dye is used to identify possible defects, cracks, or discontinuities in a seal ring.
- Bake test: In this test, the seal ring was subjected to controlled temperatures to evaluate its stability and performance under heat conditions.
- Steam aging: The seal ring was exposed to steam conditions to simulate aging and evaluate its resistance to moisture.

- Solderability test: The ability of the sealing ring to be properly soldered without affecting its properties is verified.
- Customer use test: Specific tests requested by the customer were carried out to evaluate the performance of the sealing ring under real-use conditions, such as wire placement and die attachment.
- In addition to hardness and grain size measurements in the laboratory, these additional tests address key aspects related to tightness, thermal resistance, sealing, moisture resistance, and performance in real-life conditions, ensuring the quality and reliability of the SMD sealing ring.

3.3.2.7 7D Preventive Measure

At this stage, a thorough evaluation of the SMD batch performance in processes and systems is conducted to determine the underlying causes of the problem and establish appropriate preventive measures. The most impactful defects are determined post-validation to verify the quality of the validation and how it affects product flow. For this purpose, a Pareto chart is used for defects. In addition, the post-validation batch yield trend was analyzed, and the operator profile was described to find ways to control the variables involved in the human factor.

3.3.2.8 8Ds Celebration of Achievements

During the achievement celebration phase, various activities are carried out to recognize and highlight the achievements in solving the problem. Feedback is provided to learn from the experience and apply the lessons learned in similar future projects. Generally, we seek to foster an environment of learning and collaboration that allows the entire team to feel recognized and motivated to achieve new goals in the future. Additionally, to expand knowledge to other areas and allow the flow of information to continue, a meeting is held to congratulate employees on their progress and achievements so that everyone feels pleased with the results.

3.4 RESULTS

3.4.1 Results of Stage 1: Creation of the Task Force

To ensure success and effectively address the challenges that may arise, a team was assembled with engineers from various backgrounds (see Table 3.2). Each

TABLE 3.2 Profile matrix team 8Ds

POST	*PROFILE*	*ACTIVITIES*
Process Engineer	Expert in process optimization and control to improve efficiency and quality in industrial production.	Define the operating parameters of the annealing furnace. Perform load and furnace capacity tests. Supervise the annealing process and perform quality controls. Analyze and document the results obtained during validation. Implement corrective and preventive actions in case of deviations.
Quality Engineer	Quality assurance expert to ensure conformity and continuous improvement in products and processes.	Perform sampling and quality testing of annealed products. Perform statistical analysis of data collected during validation. Evaluate the compliance of the annealing furnace with applicable standards and regulations.
Production Supervisor	Responsible for coordinating and supervising production to achieve quality, efficiency and on-time delivery objectives.	Coordinate and schedule annealing furnace validation activities. Ensure that established procedures and standards are followed during validation. Communicate and report any deviation or incident during validation. Collaborate with the engineering and quality team in the resolution of problems or adjustments needed.
Maintenance Engineer	Responsible for ensuring the availability and reliability of equipment and systems for the optimal operation of the plant.	Check and verify the condition of the annealing furnace facilities and equipment prior to validation. Collaborate in the installation of sensors and devices necessary for the validation of the furnace. Continuously monitor furnace performance indicators during validation. Provide technical support and troubleshoot any problems related to the operation of the furnace during validation.

discipline brought unique expertise and approaches that enriched the team and promoted problem-solving from different perspectives.

3.4.2 Results of Stage 2: Definition of the Problem

The problem of validating the annealing process in the Tijuana plant arises. This conflict, as mentioned earlier, arose from the lack of processing capacity, the need to increase the production level, and the reduction in material preparation time. Figure 3.4 shows the current and proposed diagrams of the annealing process. As can be observed, once the annealing was validated at the Tijuana plant, two transfer activities between the San Diego and Tijuana plants were eliminated.

The quality criteria used to validate the compliance of sealing ring annealing are as follows:

- Dimensions and tolerances: The ring has precise dimensions that allow it to fit the geometry of the sealing system properly. Measurements such as the inside diameter, outside diameter, and thickness were verified, and it was ensured that the tolerances

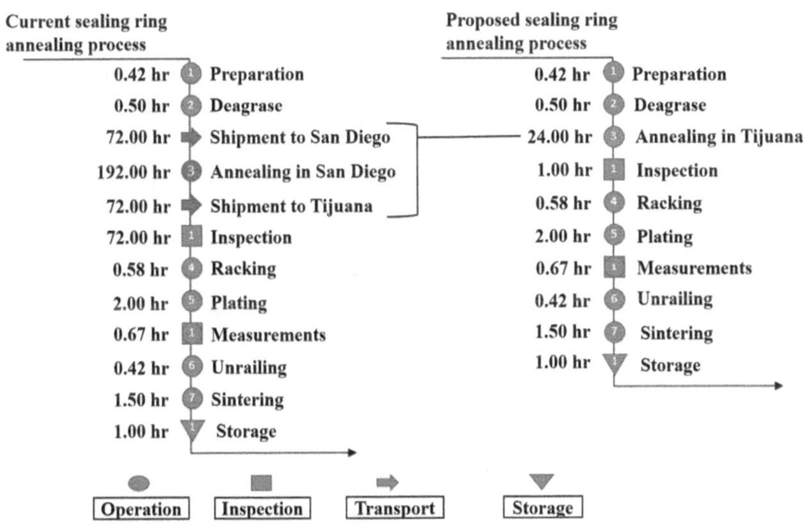

FIGURE 3.4 The current and proposed flowcharts for the sealing ring annealing process

specified by the client were met. The sealing ring had its respective nominal measurements and tolerances, which ranged between 9% and 1% of nominal value.

- Material: The sealing ring material was subjected to hardness measurements to ensure compliance with the purpose of annealing.
- Compressive strength: The ring can compress and create an effective seal without permanent deformation.
- Tightness: The ring was able to provide an effective seal to prevent the leakage of liquids, gases, or other media. Tightness tests (functional tests), such as pressure or leak tests, were performed to verify sealing capability.

The validation of the furnace for the annealing process at the Tijuana plant, along with the consequent relocation of the heat-treatment process from San Diego, involved significant changes in several related areas. The areas affected by this validation are described as follows:

- Heat treatment process: Following the validation of the furnace at the Tijuana plant, the heat treatment process will no longer be conducted at the San Diego plant. This area was directly impacted because the operation and supervision of the furnace moved to the Tijuana plant.
- Logistics and transportation: The transportation of materials from Tijuana to San Diego for heat treatment was no longer necessary, which changed the logistics and supply chain planning.
- Quality control: Quality criteria and parameters were established to ensure that the materials treated in the furnace met the required standards. These criteria are that the grain size must be greater than 38 μm and must have a hardness greater than 157 Rockwell units.
- Training and human resources: The necessary human resources were assigned to ensure a smooth and successful transition from the heat-treatment process to the plant facilities in Tijuana, requiring a preparation table and a person to place the material in its respective trays.

Validation of the annealing furnace involved verifying several critical aspects to ensure that the process was reliable and met quality requirements. Some variables considered throughout the validation were as follows:

- Temperature: The temperature was maintained at a constant level throughout the furnace, within the specified limits for the annealing process, with a maximum temperature of 1,050°C.

- Time: The retention time of the material in the furnace was kept constant and adjusted as necessary, within the range of 3–3.5 hours. This resulted in adequate annealing.
- Airflow: The airflow in the furnace was constant and uniform to ensure uniform temperature throughout the furnace. The same was true for the atmosphere, its composition to prevent damage to the material and to economize the activity. This aspect was regulated using a vacuum in the loading zone, with measurements ranging from 748 to 756 Torr.
- Material uniformity: The annealed material was uniform in composition and structure, thereby ensuring that the annealing process was effective. To control this variable, the furnace was cleaned to reduce or eliminate the formation of hydrides in the atmosphere.
- Maximum load: A maximum load of no more than 35 kg of material in the furnace was assigned, recommended for the base according to the user's manual.
- Equipment calibration: The equipment used to measure the temperature and other furnace parameters was calibrated and functioned properly.
- Safety protocols: Appropriate safety protocols were followed to ensure that the annealing process was safe for both operators and equipment.

The temperature profile obtained from the temperature and time parameters, corresponding to a peak at 1,050°C, with retention at that temperature and a gradual decrease over 3.34 hours, for subsequent ambient temperature-controlled cooling, is shown in Figure 3.5.

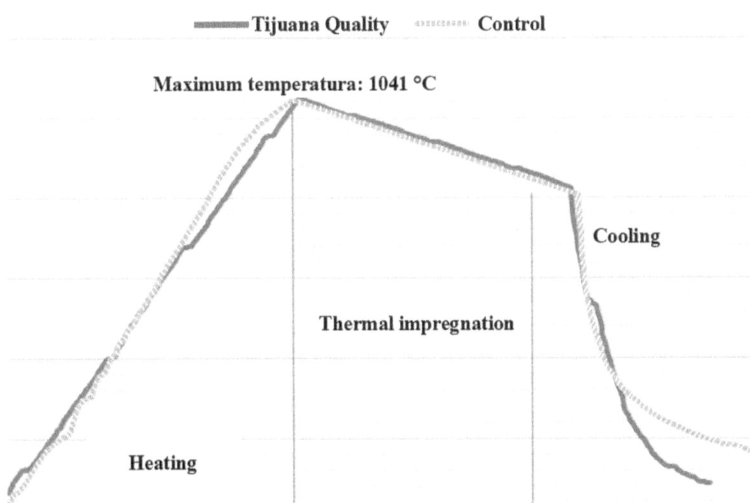

FIGURE 3.5 Temperature profile of the annealing furnace

According to the standards mentioned in Section 3.3.2.2, the aspects to be considered for the installation of the furnace correspond to the safety elements and signs.

The installation of the completed machinery included the implementation of safety signs, such as the location of the fire extinguisher, exposure to high temperatures, high voltage, danger of electrostatic discharge, and an emergency stop button for the machine. The safety elements included in the installation are thermal gloves designed for high temperatures (up to 1,000°C), waste filters on the furnace motor located at the rear of the furnace, fire extinguishers, a fume extraction hood, and a piping system. In addition, hoses identified according to their fluids, and an emergency button for a total shutdown of the equipment in the event of unexpected events that could compromise the integrity of the loader or its attachments.

The control indicators located in the annealing furnace were calibration and maintenance labels, analog display indicating the internal temperature of the annealing chamber, customs, and border crossing label, hydrogen and nitrogen inlet meter, inspector control labels, fluid inlet and outlet pressure gauges, and a final program display that provided information regarding the status of the annealing chamber for post-cooling management.

3.4.3 Results of Stage 3: Containment of the Problem

When loading and unloading material into the machine, various associated risks were identified and considered to ensure operator safety and prevent accidents. During the project, measures were taken to avoid injuries caused by trapped fingers or hands when handling the material, actions were implemented to reduce the risk of falling when transporting heavy objects, and safety protocols were established to avoid burns or cuts when interacting with the moving parts of the machine. In addition, training was provided for safe load handling, and the use of personal protective equipment was promoted.

Initially, the material loading tower, known as Tower A, which was used in the annealing process to facilitate validation in Mexico, was requested by the Tijuana plant from the San Diego plant. The tower consists of a structure made of molybdenum rods and blocks, which, using interlocking floors, can hold alumina ceramic plates containing the material in a rectangular pattern.

At the time of the initial loading of the furnace, the weight of the structure made it difficult to handle because it exceeded the 35 kg specified in the furnace user manual. This constitutes a risk for the machine because if the resistors came into contact with the metal surface due to electrical transfer, it could have caused a spark, a fire, or a shortage in the oven's electrical system.

3.4.4 Results of Stage 4: Root Cause Identification

The general problem in the containment of the problem corresponds to the tower structure, and the secondary problems are load optimization, poor material loading, and inadequate training. As can be seen in Figure 3.6, these problems are directly related to the method and labor employed.

The different points analyzed corresponded to each of the 6M. In the machinery, temperature plays a crucial role, as it must be raised to the point of change of crystallization of the grains and their level of exposure. The weak internal atmosphere is a mixture of two parts nitrogen to one part hydrogen (the actual ratio is not mentioned for confidentiality reasons).

The methods failed due to the shape and type of material load, as well as the structure used for introducing it into the furnace. The structure of Tower A represents a relatively large weight to be handled by one person. Bottlenecks, owing to material loading, represent a constant threat to the material; it is necessary to have the most significant amount of material available in the shortest possible time. Having defined the tower structure, we proceeded to make adjustments so that it could be introduced in different rounds with less weight.

The problems related to labor did not represent a failure model as such since it was a new activity where the personnel were trained as the validation

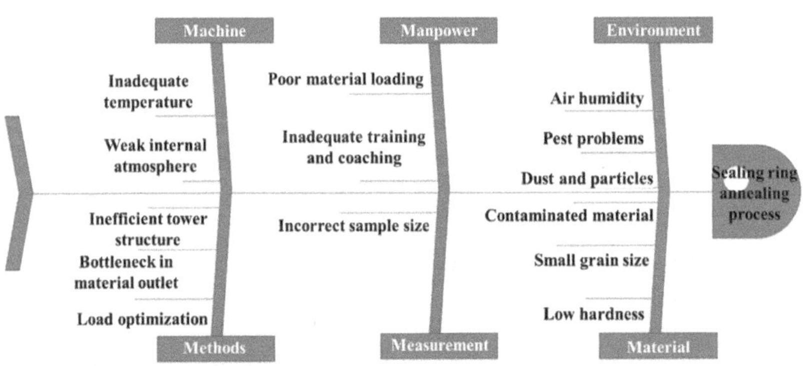

FIGURE 3.6 Ishikawa diagram of the annealing process

progressed. Poor material loading is related to the tower structure, which is understood as the arrangement of material that does not allow separation of the pieces between them and that, when passing through high temperatures, results in welding them together. This indicates a defect in the process.

The measurement in the annealing process presented failures at the moment of taking the retention samples. On two occasions, the number of samples exceeded that indicated by the Quality Department. According to a 95% confidence level and a validation batch size of 100 pieces, five samples were taken per batch, while 10 pieces were taken in two batches.

However, the environment presented failure patterns related to humidity conditions. When there is a relatively low outside temperature with a high dew point, the water droplets affect the quality of the seal and the generation of a controlled atmosphere inside. Pests, such as insects and flies, proved to be a significant contaminant; therefore, when the oven is open, constant care must be taken to keep the oven and its interior clean. Similarly, dust and particles are related to the poor quality of the sealing of the material, which is related to the previous problem. For this solution, only more cleaning was required at the beginning and end of each load.

The raw material failed due to contamination on its exterior and interior (composition), which allowed the deposition of carbon hydrides in its interior. During the annealing process, these hydrides generated a chemical reaction that prevented the correct grain enlargement of the material, which in turn affected the hardness of the material.

Analysis of the failure modes across different process categories (methods, machinery, labor, measurement, environment, and raw materials) revealed that all of them were a common cause of the lack of control and adequate follow-up of the loading parameters and procedures. This lack of control results in errors in the execution of the methods, inadequate operation of the machinery, lack of personnel training, incorrect sampling, unfavorable environmental conditions, and problems related to the quality of the raw material used.

3.4.5 Results of Stage 5: Development of Ongoing Corrective Actions

The method of loading and preparing the material was changed to solve the problem related to the weight of the structure. Toward this purpose, Tower B was introduced.

According to the demand proportions of the sealing ring batches, 100% of the production was required: 52.43% for SMD A, 34.05% for SMD B, and 13.51% for SMD C; that is, the furnace capacity could be divided into these proportions to continuously process the materials of the three parts. With a furnace charge capacity utilization efficiency of 95.65%, two batches of SMD A, one batch of SMD B, and one batch of SMD C could be charged simultaneously to facilitate the start-up.

3.4.6 Results of Stage 6: Implement and Validate Corrective Action

Tower B, with a 46-story structure, weighed 30,2623 kg and fell within the 35 kg limit. In addition, Tower B's load capacity was higher, with a maximum load of seven lots of SMD A sealing ring, two lots more than that of Tower A. This temporarily eliminated the ergonomic problem related to the manual handling of the material load and increased the furnace capacity. The change in the structure of the material-charging tower did not exhibit significant variations in the heating and cooling curves.

The data captured for the correlation analysis consisted of 530 temperature samples, one temperature data for each minute-to-minute variation, that is, minute one corresponds to a temperature of 67°C, minute 200–900°C, to compare minute 200 of the initial curve and the modified curve to observe the variation between them. Given an H_0, where there is no correlation, and an H_1, where there is a correlation, by obtaining a test statistic value of 168.57284 higher than the critical value of 1.96446, the null hypothesis H_0 is rejected at a significance level of 5%. The correlation coefficient between the temperature curves is 99.08%.

Three batches were processed for laboratory tests, including grain size analysis, microscopic inspection, and hardness tests. These three lots were accepted in all tests and were, therefore, validated. The grain sizes of the batches were similar to each other with respect to the processed batches and the control batches. In addition, one lot without heat treatment had a significantly smaller kernel size than that subjected to heat treatment.

Table 3.3 shows the ANOVA results. The mean squares of the error were lower than those of the treatments. That is, most of the variation in the data was

TABLE 3.3 ANOVA results of grain size

SOURCE OF VARIATION	*SS*	*DF*	*MS*	*F*
Treatments	1974.010445	4	493.5026112	33.8111
Error	218.93805	15	14.5959	
Total	2192.948495	19	115.4183418	

TABLE 3.4 Hardness ANOVA results

SOURCE OF VARIATION	*SS*	*DF*	*MS*	*F*
Treatments	2147.507657	3	715.8358858	0.8991
Error	9554.188442	12	796.1824	
Total	11701.6961	15	780.1130733	

due to the differences between the groups or treatments, rather than chance or experimental error.

A critical *F*-value of 3.0556 was obtained using an alpha level of 0.05. This is lower than the *F* value of 33.8111 shown in Table 3.3. Therefore, the alternative hypothesis that variations exist between treatments was accepted. When the adjusted *R* value[2] was obtained, 87.35% of the data corresponded to the control data. Regarding the multiple range tests for grain size, only lot 1 was significantly equal in grain size to the control lot, but all lots had a grain size greater than 38 μm; therefore, all were accepted under this quality criterion. In contrast, the LSD results indicated that all lots varied significantly with the lot that did not undergo annealing.

Similarly, concerning the batch hardness results, a value of $\alpha = 0.05$, with a value of 3.4903, is obtained. This value was higher than the *F* value of 0.8991, as shown in Table 3.4. Therefore, the null hypothesis of no variation between the treatments was not rejected. When the adjusted R^2 value was obtained, it was determined that −2.06% of the data corresponded to the control data. However, the results of the multiple-range tests for hardness indicated no difference between treatments. That is, all lots were statistically equal, which was verified by the mean sum of the treatments and errors. Because the error was higher, the variation between data came from the samples, not the treatments. In addition, there was no difference when calculating the LSD between the batches.

Finally, within the engineering verifications, the batches passed the visual tests, using the SEM to verify the appearance of the grains in the treatments. Each batch was visually acceptable and had a uniform crystalline structure without the formation of carbon hydrides (contaminants) on the surfaces of the parts. All batches passed all quality tests in the SMD assembly; all were accepted and validated.

3.4.7 Results of Stage 7: Preventive Measure

Several potential failure modes that could affect the process quality and annealing results were identified during furnace validation. Some of these failure modes are improper calibration of the temperature sensors, inadequate ETs, unstable or deviated temperatures, improper atmosphere in the furnace, and possible equipment integrity failures. These failure modes were considered during the validation, and control measures were implemented to prevent them and ensure a successful annealing process.

In addition, proper calibration of the temperature sensors was performed to ensure accurate measurement. The ETs were carefully controlled to avoid under- or over-annealing. Adjustments were made to the furnace temperature to ensure uniformity of annealing. Similarly, visual control measures were implemented on the pressure gauges to ensure an adequate atmosphere in the furnace. Periodic inspections were conducted monthly to detect any flaws in the integrity of the equipment, and corrective actions were taken promptly. These measures mitigated potential failure modes and ensured the efficiency and reliability of the furnace validation process.

The defects changed throughout the validation, as shown in Figure 3.7, from an initial number of 5,185 to 3,736. Although the number of defects was reduced, the number was still considered relatively high. It can be interpreted that while reducing the number of defects due to the exposed metal and damage, others increased, as was the case for discoloration and bridging. Some possible causes of discoloration include exposure to high temperatures during annealing, chemical environmental reactions, or contaminants in the process. Bridging, on the other hand, could result from an incorrect distribution of the material during fabrication or inadequate annealing, which generates an

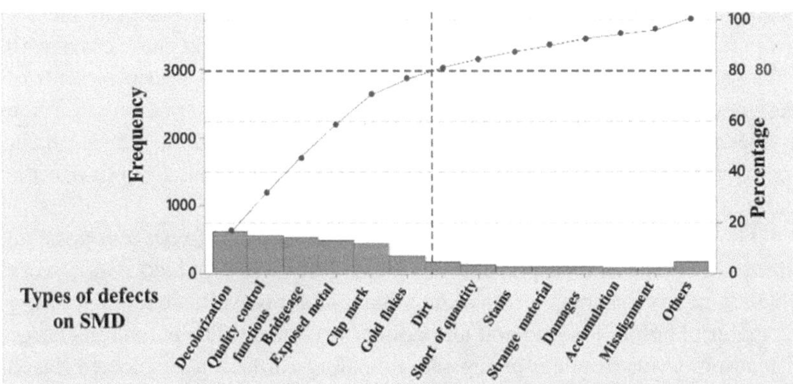

Frequency	618	548	522	490	439	241	157	124	106	100	93	80	65	154
Percentage	17	15	14	13	12	6	4	3	3	3	2	2	2	4
Cumulative percentage	17	31	45	58	70	76	81	84	87	90	92	94	96	100

FIGURE 3.7 SMD defect Pareto after validation

excess of fusion and bonding between the parts of the ring. Further tests were planned to be carried out during the following months. However, the plots' yields showed a nonsignificant improvement of 0.48% compared to the pre-validation plots.

Annealing entailed various risks associated with personnel performing the task. Common risks, controlled by safety labels and protective equipment, included the provision of heat-resistant gloves. Therefore, personnel were instructed on proper use. In addition, adjustments were made to the ventilation system to ensure the adequate removal of toxic gases and vapors, thereby protecting the workers' respiratory health.

Ergonomic techniques were implemented to avoid load handling injuries, and appropriate tools were provided to safely handle heavy loads. The initial tower (Tower A) was replaced with a modular type by floors (Tower B) to avoid a single 30 kg load and distribute it into several loads of three to five kilograms. In addition, personnel were trained in best practices to avoid musculoskeletal injuries. Equipment inspections were conducted in the area of electrical safety, and necessary repairs and maintenance were performed. Electrical connection and disconnection procedures were reinforced, and personnel were trained on electrical hazards and corresponding preventive measures. A rigorous fire and explosion prevention plan was established. Fire extinguishers were installed strategically in the annealing area for quick action.

On the other hand, the personnel trained to execute the activity were delimited into two parts: a packing and unpacking operator who accommodates the sealing ring on the Tower B structure, which can be loaded in segments by a technician or engineer in the annealing furnace. A clear and accurate picture of operations was obtained through direct observation and interaction with employees, identifying opportunities for improvement and possible points of inefficiency. This approach, based on "*Gemba*" or site observation (from the Japanese), allowed a complete understanding of the workflow, the identification of waste, and the implementation of practical solutions to optimize the sealing ring preparation process.

The workstation represents a point to be considered to prevent possible failures. The operator responsible for the activity was provided with a work table 1.7 meters long by 1 meter wide, a chair, and trays suitable for processing the material before loading and unloading the furnace. The person in charge of loading, unloading, and using the annealing furnace was selected based on their technical knowledge and skills in handling loads, safety awareness, organizational ability, ability to follow instructions, quality orientation, and teamwork skills. Their activities include planning and coordinating loading and unloading, following precise instructions, ensuring product quality,

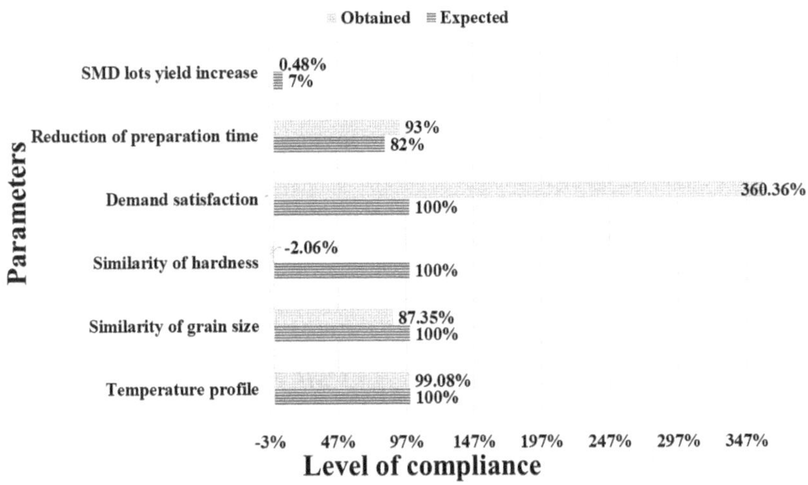

FIGURE 3.8 Results obtained for each variable

and collaborating with other team members. The work team's achievements correspond to the different aspects analyzed throughout the project, including temperature, grain size, hardness, demand, time, and yield, as shown in Figure 3.8.

As part of the final validation stages, personnel training requires documentation for subsequent control and follow-up in quality programs and external inspections. For this purpose, two work instructions were developed, pending accreditation by the Quality Department for preparing, loading, and unloading materials from the furnace. These work instructions were meticulously designed, considering their structure and distinctive features, to ensure the optimal and safe operation of the furnace. The wording of the work instructions was carefully selected to provide clear and precise guidance to operators, ensuring that they followed the necessary steps to achieve successful results.

The waste generated by the process is not considered hazardous or a control material; therefore, it is deposited in conventional trash containers (excluding plastics, paper, and cardboard). Oven calibration is not required according to the user manual, but a preventive maintenance manual is planned every three months. Similarly, a monthly review of the installations is performed to accelerate or delay maintenance according to the conditions of the labels, hoses, pipes, and pressure gauges.

3.4.8 Results of Stage 8: Celebration of Achievements

All team members were rewarded with an in-house meal and recognition of individual and team achievements, as well as job promotions. Throughout the validation process, there was a process of emerging problem-solving, in which everyone focused on releasing the problem and allowing the process to flow.

3.5 CONCLUSION

The general objective of this project was to validate, using the 8Ds methodology, the annealing process for SMD packages of the TO department. It is concluded that this objective was achieved because the implemented methodology allowed for the correct development of the annealing process and its acceptance by the Quality Department. Subsequently, the process should be validated at the customer level when sufficient production batches are available to demonstrate the sustainability of validation.

The characteristics of the annealing process are statistically equal to or similar to those of the desired or perfect control process. The heating and cooling curves were significantly equal at 99.08% according to the applied Pearson's correlation coefficient. The grain size between the test batches was equal at 87.35%, which is the statistic that differs the most from the control batch. The visual appearance of the lots was uniform, with three similar sizes, ranging from 49 μm in length. Finally, the hardness achieved after the heat treatment was significantly equal at –2.06%. This negative value indicates that the range of variation originates from the data between samples and not between batches, and more batches are necessary to prove the contrary.

In addition to implementing accredited validation in the company, production levels increased. Regarding the first specific objective of supplying 100% of the demand for the sealing ring, the sealing ring was processed at relatively low levels, given the preparation time required when it was sent to the San Diego plant for annealing and then sent back to the Tijuana plant. Currently, 4800 pieces of sealing rings can be processed per day as a stable load that regulates the level of demand. This means that 5.44 times more sealing rings can be processed than before. The production level was 444.26%. Considering the increase in demand level from 881 to 1332 sealing rings, the current process can satisfy 360.36% of the demand.

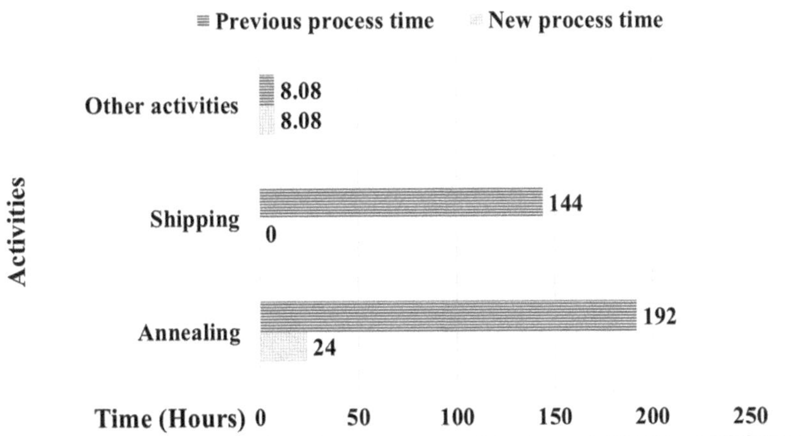

	Other activities	Shipping	Annealing	Total
New process time	8.08	0	24.00	32.08
Previous process time	8.08	144.00	192.00	344.08

FIGURE 3.9 Sealing ring process time before and after validation

The second objective was to reduce the preparation process time for ring sealing by at least 82%. It was concluded that this objective was successfully achieved, resulting in a 93% reduction in the preparation time. With this, the process went from 344.08 to 24.08 hours. As shown in Figure 3.9, the annealing time was reduced by 88%, the shipping time was eliminated, and the rest of the activities were maintained at normal levels. The reduction in the waiting time between shipment, treatment, and return of the batch allowed the processing of a larger quantity of sealing rings and, therefore, to have a greater availability of material for SMD assembly.

The third objective was to increase the batch yield by 7% for the SMD packages. Thus, it was concluded that this objective could not be achieved. As can be seen in Figure 3.10, the yield of the batches that were assembled after furnace validation did not show a significant increase, with a rise of only 0.48%. The yield of good parts per batch went from 91.41% to 91.89%. Defects related to the exposed metal and damage were reduced by 10.6% with respect to the annealing process at the San Diego plant; however, defects such as click marks, discoloration, and bridging increased by 11.7%. It was not possible to eliminate the defects to be treated, and others arose that had not been considered before, for which it will be necessary to analyze their causes and consequences.

The choice of the 8Ds methodology to address the furnace validation project proved to be highly effective and appropriate. This methodology proved

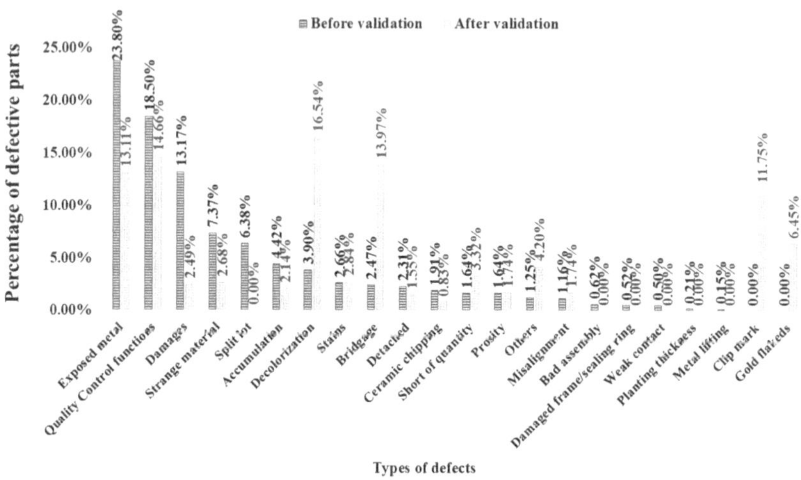

FIGURE 3.10 SMD defects before and after validation

to be a robust and structured tool that enabled the identification, analysis, and resolution of the problems encountered during the SMD sealing ring annealing process. The eight-step sequence facilitated a systematic and comprehensive approach to root cause analysis and corrective action. In addition, it fostered collaboration and teamwork, which involved diverse team members in decision-making and solution implementation. The 8D methodology was the best choice due to its structured approach, ability to address complex problems, and its contribution to the continuous improvement of the furnace validation process.

The results of this project align with those obtained in other cases where the 8D methodology was used. For example, Realyvásquez-Vargas et al. (2020) applied this methodology to reduce defects in the cable assemblies of motors by 75%. Similarly, Zhou et al. (2016) proposed a procedure for quality improvement based on 8Ds methodology. Cheng and Chen (2010) applied this methodology to identify and correct the causes of nonconformity when a customer filed a complaint about a product or service. Consequently, these authors decreased the defect rate from 28% to 0.5%, which translated into profits of at least $22 million. Finally, they conducted a study in a spring and at a stamping parts manufacturing company, whose process produced defective parts at a rate of 17.07%. With the application of 8D methodology, the defect rate was reduced to 4.91%.

With this, it is demonstrated and confirmed that the 8Ds methodology is an excellent strategy to reduce defects and customer complaints, and to validate production processes.

REFERENCES

Ahir, Keyur B., Khushboo D. Singh, Sushma P. Yadav, Hetal S. Patel, and Chetan B. Poyahari. 2014. "Overview of Validation and Basic Concepts of Process Validation." *Scholars Academic Journal of Pharmacy* 3 (2): 178–90.

AsociaciónMexicana de Constructores de Invernaderos. 2013. "Norma Mexicana Para El Diseño y Construcción de Invernaderos. (NMX-E-255-CNCP-2008)." https://es.scribd.com/doc/157292212/NMX-E-255-CNCP-2008-Resumen-080310.

CámaraMexicana de la Industria de la Construcción. 2023. "Normas Oficiales Mexicanas." https://www.cmic.org.mx/comisiones/Sectoriales/medioambiente/Varios/Leyes_y_Normas_SEMARNAT/NOM/nom.htm.

Cheng, Hsiang-Ru, and Bor-Wen Chen. 2010. "A Case Study in Solving Customer Complaints Based on the 8Ds Method and Kano Model." *Journal of the Chinese Institute of Industrial Engineers* 27 (5): 339–50. https://doi.org/10.1080/10170669.2010.495508.

DOF—Diario Oficial de la Federación. 2008. "NORMA Oficial Mexicana NOM-026-STPS-2008, Colores y Señales de Seguridad e Higiene, e Identificación de Riesgos Por Fluidos Conducidos En Tuberías." https://www.dof.gob.mx/normasOficiales/3541/stps.htm.

DOF—Diario Oficial de la Federación. 2009. "NORMA Oficial Mexicana NOM-030-STPS-2009, Servicios Preventivos de Seguridad y Salud En El Trabajo-Funciones y Actividades." https://www.dof.gob.mx/normasOficiales/3923/stps/stps.htm%23:~:text%3DNORMA%2520Oficial%2520Mexicana%2520NOM%252D030%2Cel%2520trabajo%252DFunciones%2520y%2520actividades%26text%3DEstablecer%2520las%2520funciones%2520y%2520actividades%2Caccidentes%2520y%2520enfermedades%2520de%2520trabajo.

DOF—Diario Oficial de la Federación. 2010. "NORMA Oficial Mexicana NOM-002-STPS-2010, Condiciones de Seguridad-Prevención y Protección Contra Incendios En Los Centros de Trabajo." https://www.dof.gob.mx/normasOficiales/4228/stps/stps.htm.

DOF—Diario Oficial de la Federación. 2012. "NORMA Oficial Mexicana NOM-001-SEDE-2012, Instalaciones Eléctricas (Utilización)." https://dof.gob.mx/nota_detalle_popup.php?codigo=5280607.

Lefeber, E., R. A. van den Berg, and J. E. Rooda. 2004. "Modeling, Validation and Control of Manufacturing Systems." In *Proceeding of the 2004 American Control Conference*, 4588. Boston: Institute of Electrical and Electronics Engineers (IEEE). https://doi.org/10.23919/ACC.2004.1384033.

Psomas, Evangelos L., Dimitrios P. Kafetzopoulos, and Christos V. Fotopoulos. 2013. "Developing and Validating a Measurement Instrument of ISO 9001 Effectiveness in Food Manufacturing SMEs." *Journal of Manufacturing Technology Management* 24 (1): 52–77. https://doi.org/10.1108/17410381311287481.

Realyvásquez-Vargas, Arturo, Karina Cecilia Arredondo-Soto, Jorge Luis García-Alcaraz, and Emilio Jiménez-Macías. 2020. "Improving a Manufacturing Process Using the 8Ds Method. A Case Study in a Manufacturing Company." *Applied Sciences* 10 (7): 2433–58. https://doi.org/10.3390/app10072433.

Shabi, Jacob, Yoram Reich, and Roee Diamant. 2017. "Planning the Verification, Validation, and Testing Process: A Case Study Demonstrating a Decision Support Model." *Journal of Engineering Design* 28 (3): 171–204. https://doi.org/10.1080/09544828.2016.1274964.

Tripathi, Varun, Somnath Chattopadhyaya, Alok K. Mukhopadhyay, Shubham Sharma, Jujhar Singh, Danil Yurievich Pimenov, and Khaled Giasin. 2021. "An Innovative Agile Model of Smart Lean–Green Approach for Sustainability Enhancement in Industry 4.0." *Journal of Open Innovation: Technology, Market, and Complexity* 7 (4): 215. https://doi.org/10.3390/JOITMC7040215.

Zhou, Fuli, Xu Wang, Tshepo Mpshe, Yi Zhang, and Yuhang Yang. 2016. "Quality Improvement Procedure (QIP) Based on 8D and Six Sigma Pilot Programs in Automotive Industry." In *Proceedings of the First International Conference Economic and Business Management*, edited by W.C. Du, 275–81. Qingdao, China: Atlantis Press. https://doi.org/10.2991/FEBM-16.2016.43.

Case Study 2. Validation of Ultrasonic Sealer for Laparoscopy Model Production

4

4.1 INTRODUCTION

Validation plays a critical role in the development, installation, and use of these devices in the medical industry (Xu et al. 2018). Validation ensures that equipment and devices meet the functional, reliability, and quality requirements of both customer and regulatory authorities (Xu et al. 2018). Equipment validation has become a regulatory requirement for pharmaceutical companies that must precede the validation of new equipment and instruments (Jindal et al. 2020). Literature and websites present information on various regulatory agencies, professional societies, seminars, symposia, and professional consultants for equipment validation

DOI: 10.1201/9781003650966-4

(Dwivedi 2006). Equipment validation included the following studies (Dwivedi 2006):

- Design Qualification (DQ)
- Installation Qualification (IQ)
- Operational Qualification (OQ)
- Performance Qualification (PQ)

Equipment qualification studies should be performed for all process equipment (fermenters/bioreactors, chromatography, microfiltration/ultrafiltration systems, tanks and vessels, autoclaves, CIP and SIP systems, lyophilizers, to name a few); utility equipment (WFI, clean steam, solvent supply, among others); supply equipment and accessories (gases, filters, raw materials); automated systems (computers, DCS, PLC); and critical facility equipment (HVAC, warehouses, storage chambers, shipping containers, etc.) (Dwivedi 2006).

The literature presents several case studies on the validation of the equipment, devices, and software in the medical industry. For example, Kaivosoja (2015) validated a software that guides the production of medical devices and collects traceability data, including device history files. The author mentions that the validation process must comply with the guidelines and regulations of the FDA guidelines and regulations. Finally, software validation results in increased usability, efficiency, and flexibility. Pluta et al. (2010) presented an analysis of a practical validation case that includes equipment cleaning and inspection of cleaned equipment. In this case, the authors mentioned that although the equipment had been cleaned using a validated cleaning procedure, manufacturing residues were observed on a specific piece of equipment after cleaning. The following three issues were investigated: equipment cleaning assessment, equipment cleaning, and cleaning validation. The analysis determined that the root cause was inadequate rinsing of the equipment owing to undefined steps in the cleaning procedure. Corrective actions were implemented to verify and maintain the validation process.

Finally, Fields (2010) presented a case study on equipment validation in a pharmaceutical packaging facility. The event involved a review of the documentation associated with the qualification of a cartoning machine. This review demonstrated a serious lack of understanding of fundamental validation and qualification practices and requirements, including protocol preparation, data handling, and approval practices. To address this problem, the author cites specific GMP requirements, suggests corrective and preventive actions, and includes specific equipment qualification activities, basic validation and qualification procedures, research procedures, training, data handling, and other relevant practices.

This background clarifies that equipment validation in various industrial sectors, including manufacturing, plays a fundamental role in process performance and, in the case of manufacturing, in product quality, production times, and costs. The following sections present a case study of the validation of an ultrasonic sealer for the production of a laparoscopy model. Additionally, the objectives, applied methodology, and results are described.

4.2 CASE STUDY

The present study was conducted at a leading medical device manufacturing company. This company complies with certifications, regulations, and strict quality standards, including ISO 13485 and the FDA guidelines. In addition, the company has a global presence in 150 countries, with over 350 locations worldwide. The case study was conducted on one of the two plants located in Tijuana, Mexico. This plant comprises more than 5,000 employees, who are led by the manufacturing director, along with his management team. This team was formed by the engineering director, maintenance manager, engineering manager, and Senior engineering manager, who is the project manager's direct boss.

The company is a medical device manufacturer and manufactures various products, including surgical pencils, adhesive panels for energy output, food pouches, and oximeters.

The plant consists of three main buildings in the house production areas. It also had a warehouse, two canteens, two parking lots, and an administrative area.

This project is conducted in the project department, which is led by the project manager, who has an engineering team composed of four engineers, a project coordinator, a project technician, and a project intern. This department is responsible for carrying out new product introductions and transfer projects. Currently, this team does not have a designated work area; therefore, it can operate in any production area of the plant that requires it. The project department is integrated by a Project Manager, a Project engineer, a Senior project engineer, two Junior project engineers, a project coordinator, a project intern, and a project technician. The activities of each position in the department are as follows.

- Project Manager: Responsible for the coordination and delegation of project activities.
- Project engineer: He is in charge of the validation execution of the project transfer. He was also in charge of reviewing and approving documents and changes related to the project.

- Project Engineering Intern: Provides support to the transfer project with activities, such as documentation and execution of validation through IQ, operation qualification (OQ), and process performance verification (PQ). In addition, he provides support for the production lines during the transfer run under risk. He was responsible for the creation and release of documents and changes related to the transfer.

More specifically, the project was conducted in the surgical pencil area. This area comprises various manufacturing lines. The project will be carried out in the accessory line, where the ultrasonic sealing process is performed using an ultrasonic sealer. Figure 4.1A shows the layout of the surgical pencil area, while Figure 4.1B shows the layout of the accessories line.

4.2.1 Problem Statement

As part of the company's procedure, it is necessary to validate all equipment, whether new or transferred from another plant of the company. Thus, it was demonstrated that the equipment was capable of producing the desired product. Ultrasonic sealer is a piece of equipment transferred from the Colorado plant in the United States to the Tijuana plant in Mexico. This equipment belongs to the "manual laparoscopy unit" process line. For this equipment to be used during the process, validation must be performed in which the equipment complies with a Cpk index greater than or equal to 1.33, which is a requirement demanded by the company's internal validation procedure. Likewise, it is necessary to determine the optimal parameters for the equipment to offer its best performance through a DOE; otherwise, it would not be possible to ensure that the equipment is manufacturing high-quality products.

The production line consists of four operations, as listed in Table 4.1. As can be seen, the operation time in the ultrasonic sealing process is 27.31

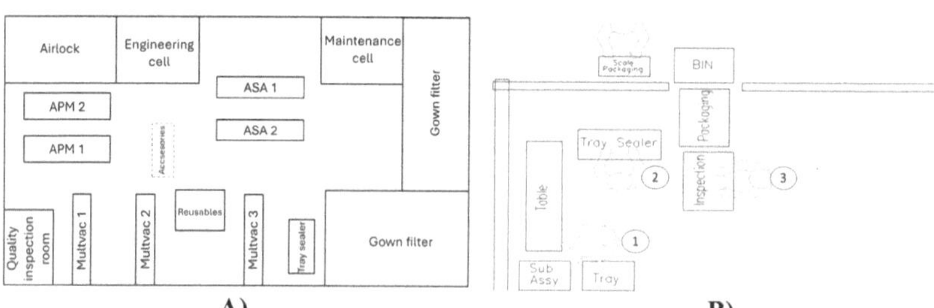

FIGURE 4.1 A) Surgical pencil area distribution, B) Accessory line layout

TABLE 4.1 Breakdown of operations of the accessories line

OPERATION	*DESCRIPTION*	*TIME (SECONDS)*	*NUMBER OF OPERATORS*	*PIECES/ HOUR*
1	Pencil and Tray Inspection and Cleaning	25.87	1	132
2	Tray and Lidstock sealing	27.31	1	
3	Sealing Inspection	26.60	1	
4	Final packaging and placement of Box Label	26.96	1	
Total		106.74	4	

seconds, which is the highest time of the four operations; therefore, it is taken as the cycle time. Under these conditions, there was a capacity of 132 pieces per hour, estimating a weekly demand of 5,346 pieces with an average of 8.1 hours of work. Once weekly production demand is met, line balancing and improvement proposals should be made to increase weekly production by at least 10%. To reach this 10%, we must analyze whether the operator is using the necessary tools to perform their job, as well as the ergonomics at their workstation, to facilitate the flow of the line.

4.2.2 Objectives

The general objective of this project is to validate ultrasonic sealer equipment through IQ, OQ, and PQ validation by applying the design of experiments methodology to meet the internal requirements of the company. On the other hand, the specific objectives are as follows:

- Obtain Cpk ≥ 1.33 during the PQ of the ultrasonic sealer.
- Determine optimal equipment parameters during DOE execution.

4.3 MATERIALS AND METHODS

4.3.1 Materials

The following software was used for the development of this project: Agile, Minitab, Microsoft Word, and Microsoft Excel. In addition, tools such as a checklist for the verification of the installation of the equipment, a multimeter, and a pressure gauge are used. Among these materials, the components are used to produce surgical pencils. In addition, the following equipment is also used: an ultrasonic sealer, an electrical resistance tester, a Hipot, a test stand (Chatillon sealing voltage), a load measurement device (Chatillon welding voltage), and a calibrator (rocker inspection).

4.3.2 Method

The following is a detailed description of each stage of the method and its corresponding steps.

4.3.2.1 Stage 1 Development of Equipment IQ

4.3.2.1.1 Verification of Equipment Utility

The verification of the equipment utility consists of a plan by the internal company procedure, with its respective requirements, according to the utility. In addition, it comprises a report generated by the equipment. The report includes the results, the verification method (Pass/Fail), the instrument identification, the calibration due date, and the name of the person who verifies it. This report is used to verify that the plan is being complied with. During the development of this activity, various tools are used, including a multimeter, to measure the voltage, frequency, and electrical current of the equipment, and to ensure that these values fall within the acceptable range. For voltage, the acceptable values are 100–120 VAC (alternating current voltage); for frequency, 50–60 Hz; and for electric current, less than or equal to 14 Amp. Otherwise, the equipment is unsuitable for operation during the process. A pressure gauge was used to measure the downstream compressed-air pressure. This measurement must also be within the acceptable range, which is 35 PSI to 100 PSI; otherwise, the equipment is unsuitable for operation during the process.

4.3.2.1.2 Equipment Calibration Requirements

During the development of Stage 1, the equipment calibration requirements must be fulfilled and recorded in a checklist named "Equipment calibration requirements." This checklist includes two sections: Plan and Report. Similarly, the plan comprises the instrument name and the calibration range of operating equipment. On the other hand, the report includes the calibration range, the calibration identification, the calibration due date, and the name of the person who verifies it. When verified and approved by a trained person according to the area, this checklist confirms that the equipment calibration requirements are acceptable. During this activity, a pressure gauge was used to measure and ensure that the equipment calibration was within the range of 0–100 PSI.

4.3.2.1.3 Verification of Equipment Operation

During the development of this verification, the equipment performance checklist tool is used, where the "pass" or "fail" box is checked, depending on whether or not the equipment complies with the test. As with the other checklists, this one comprises two sections: Plan and Report. The plan includes the number of tests (Test #), the objective of the test, the expected result, and the verification method. The report section includes the result, the decision Pass/Fail, and the name of the person who verifies it.

4.3.2.2 Stage 2: Development of the Technical Engineering Evaluation

During this stage, an engineering technical evaluation is performed to assess the DOE performance and variation transmission analysis (VTA) for the ultrasonic sealer, which is assembled according to procedure DC00126657 "Manufacturing Instructions for Assembly of Hand Switch E2750V01," with the current revision in the Agile system.

4.3.2.2.1 DOE Execution

The points to be followed during the technical engineering evaluation (TEE) are as follows:

- The execution of this study follows the DOE Run Table generated by the Minitab® software.
- Six samples were taken for fractional factorial DOE per run for attributes and variables to perform the inspection.
- Six samples were taken for the DOE of the response surface per run for attributes and variables to perform the inspection.
- Thirty samples are taken for VTA confirmation at low parameters to demonstrate compliance/confidence at 90% compliance and 95% confidence (sample size per "Process Validation Work Instruction" procedure).
- Thirty samples are taken for VTA confirmation at high parameters to demonstrate compliance/confidence at 90% compliance and 95% confidence (sample size per "Process Validation Work Instruction" procedure).

At each parameter run, all units are inspected by the Quality Control department according to *Boulder*'s procedure "Verification Procedure E2750," and the inspection results of the quality equipment are documented.

The results of each run were documented and analyzed to determine the optimum process parameters using fractional factorial DOE, DOE response surface analysis, and VTA to obtain a proven acceptable range for the ultrasonic sealing process using an ultrasonic sealer.

4.3.2.2.2 Definition of Critical Points for Quality
The critical points to be evaluated by the Quality Control department to ensure good quality of the finished product are the following:

- Sealing integrity strength (front): It must be ≥ 6 Lb/f.
- Sealing integrity strength (back): It must be ≥ 10 Lb/f.
- Visual (aesthetic seal integrity finish): It passes, or it fails.
- Slot width: 0.201″ (visual Pass/Fail).
- Hypot: It passes, or it fails.

If any of the samples do not meet these specifications, it is not accepted by the Quality Control department.

4.3.2.2.3 Definition of Parameter Range
For the ultrasonic sealing process in the ultrasonic sealer, the following ranges were used:

- Sealing time: 100–250 milliseconds.
- Waiting time: 1–10 seconds.
- Force: 15–200 pounds.
- Pressure: 20–80 psi.
- Downstream speed: 0–9.

These are the same data obtained in the documentation shared by the plant in Colorado, USA.

The DOE results are documented in the Stage 2 Results section.

4.3.2.3 Stage 3: Equipment Operation Qualification

The purpose of the OQ plan is to provide objective evidence that the ultrasonic sealing process for the accessory line produces pencil subassemblies that meet the preestablished requirements for the process.

The objective was to develop an OQ to challenge the parameters found in the TEE of the ultrasonic sealer that is used in the ultrasonic sealing process on the accessory models in the production line. This allows the production of high-quality parts and ensures their ability to operate consistently within established limits.

4.3.2.3.1 Sample Size and Acceptance Criteria, Execution of Equipment for Sample Production, and Submission of Samples to the Required Tests

During the execution of OQ, the size of the sample of units to be manufactured should be determined. Tables 4.2 and 4.3 should be used as a guide in which the acceptance criteria for attribute data and variable data are observed. This is used to determine the type of data to be evaluated, as shown in Table 4.4. During the execution of the OQ, six criteria were evaluated, as listed in Table 4.5. These criteria were evaluated by the quality team once the units were manufactured with the help of ultrasonic sealer.

First, the severity range was measured to determine the sample size. These data were obtained from the company's internal procedures. Then, the severity range provides a confidence percentage and a conformation percentage, and finally, the sample size to perform the evaluation. Once these data are obtained and the sample size for each test is available, the required number of samples is manufactured using the parameters obtained from the DOE result. The samples were manufactured by the same engineering team, with the help of an operator. These samples were then delivered to the Quality Control department. When they finished the tests, the checklist in Table 4.4 was filled

TABLE 4.2 Minimum sample size for the OQ using the binomial approach and the severity level

RISK LEVEL	*SEVERITY RANGE 1–10*	*% OF CONFIDENCE*	*% OF COMPLIANCE*	*MINIMUM SAMPLE SIZE (N)*
Insignificant	1–6	95%	93%	42
Menor	7	95%	93%	42
Mayor	8	95%	96%	74
Critique	9	95%	98%	149
Catastrophic	10	95%	99%	299

TABLE 4.3 Minimum sample size for PQ using the binomial approach and level of severity

RISK LEVEL	*SEVERITY RANGE 1–10*	*ε/S*	*MINIMUM SAMPLE SIZE (N)*
Negligible	1–6	25%	62
Minor	7	25%	62
Mayor	8	20%	97
Critique	9	17%	133
Catastrophic	10	15%	171

TABLE 4.4 Equipment OQ and PQ checklist

TEST #	*PROCESS OUTPUT REQUIREMENTS AND SOURCE*	*TEST METHOD AND TMV*	*RISK LEVEL*	*CONFIDENCE COMPLIANCE*	*DATA TYPE*	*PLANNED ANALYSIS METHOD (ONLY VARIABLE DATA)*	*SAMPLE SIZE*	*TEST ACCEPTANCE CRITERIA*
	Visual inspection of surgical pencils							
	High Voltage Breakdown (Hi-Pot) Specification: 5.2 KV							
	Weld integrity strength (front)							
	Weld integrity strength (back)							
	Rocker arm slot width Test specification: 0.201″ (gate)							
	Continuity test and air test cycle of surgical pencils							

with the results. If any of the tests fail, it is necessary to repeat the procedure with different parameters because the equipment would still not be in optimal operating conditions.

4.3.2.4 Stage 4. Team Performance Appraisal

The purpose of this stage is to establish, through objective evidence, that the ultrasonic sealing process, using the ultrasonic sealer, consistently produces products that meet preestablished requirements. To reach this stage, it is necessary to conclude with the IQ, TEE, and OQ. Once the OQ results are obtained and acceptable, it can be assured that the equipment is capable of producing products that meet the established requirements. However, it is necessary to verify that the process, together with the equipment, can produce products that meet established requirements. For this, it is necessary to perform runs with the complete production line, and once the product is obtained, to perform tests with the help of the quality department.

4.3.2.4.1 Sample Size and Acceptance Criteria, Execution of Production Line Run for Sample Preparation, and Submission of Samples to the Required Tests

During the qualification of the performance of the equipment, the same method is used to determine sample sizes and acceptance criteria that are displayed in Tables 4.2 and 4.3, but in this case, Table 4.2 is no longer used to use Table 4.5. During the execution of PQ, six criteria were evaluated, as shown in Table 4.5. These criteria were evaluated by the quality team once the specimens were fabricated on the laparoscopy hand unit line using the ultrasonic sealer.

Once the quality team finishes the corresponding tests, the checklist in Table 4.4 is filled with the results. If any of the tests fail, the process is not suitable for operation because it does not meet preestablished requirements.

TABLE 4.5 Minimum sample size for the performance rating (PQ) using the binomial approach and the severity level

RISK LEVEL	*SEVERITY RANGE 1–10*	*% OF CONFIDENCE*	*% OF CONFOR-MATION*	*SAMPLE SIZE (N)*	*RISK LEVEL*
Insignificant	1	1–6	99%	93%	64
Minor	2	7	99%	93%	64
Mayor	3	8	99%	96%	113
Critique	4	9	99%	98%	228
Catastrophic	5	10	99%	99%	459

4.4 RESULTS

4.4.1 Results of Stage 1: Equipment IQ

4.4.1.1 Equipment Utility Verification

During the development of this activity, various utilities were evaluated, such as the voltage that the equipment required for its operation, the electrical frequency necessary to operate, the electrical current required by the equipment, and the compressed air pressure. With the use of a multimeter, the voltage was measured and the result was 117.1 VAC, which indicates that it passed the test, as the test requirement was between 100 and 120 VAC. In addition, the frequency and electrical current of the equipment were evaluated. For this, a multimeter was used, obtaining a result of 60 Hz for the electrical frequency, 50–60 Hz for the acceptable range required, and a result of 1.65 Amp for the electrical current, which is within the requirement of less than or equal to 14 Amp. Thus, it was confirmed that the equipment passed the test with respect to voltage, frequency, and electric current.

Finally, the compressed air pressure was evaluated. A pressure manometer was used to obtain a result of 94 PSI. Thus, it was verified that all utilities were within the ranges required by the company. Table 4.6 shows the results of the verification of the utility of the equipment, where the box "Results" column the box "passes" or "fails" was marked with a check mark, depending on whether or not it complied with what was specified in the "Requirements" column.

4.4.1.2 Equipment Calibration Requirements

It was verified that the pressure manometer of the ultrasonic sealer was within the calibration range required by the company. Calibration was successfully performed using a pressure gauge, and 94 PSI values were obtained. This is documented in Table 4.7, with calibration number PM-165-CD.

4.4.1.3 Verification of Equipment Operation

Table 4.8 documents the results of the equipment operation verification. This verification was performed visually, ensuring that the equipment had all of its functions enabled. If none of these functions were enabled or damaged, it did not pass the verification. In the "Pass/Fail" column, the "Pass" or "Fail" box was selected, depending on whether or not it complied with what was specified

TABLE 4.6 Equipment usability check results

PLAN		*REPORT*					
UTILITY	*REQUIREMENT(S)*	*RESULTS*		*VERIFICATION METHOD*	*INSTRUMENT IDENTIFICATION*	*CALIBRATION DUE DATE*	*VERIFIED BY*
Voltage	100–120 VAC	117.1 VAC	☒ Pass ☐ Fail	Multimeter	MM-552-CD	10/19/23	Technician
Frequency	50/60 Hz	60 Hz	☒ Pass ☐ Fail	Multimeter	MM-552-CD	10/19/23	Technician
Current	≤ 14 Amp	1.65 A	☒ Pass ☐ Fail	Multimeter	MM-552-CD	10/19/23	Technician
Compressed air	35–100 PSI	94 PSI	☒ Pass ☐ Fail	Pressure gauge	PM-1596-CD	10/22/23	Technician

TABLE 4.7 Results of equipment calibration requirements

PLAN		*REPORT*			
Instrument name	Calibration range of operating equipment	Calibration range	Calibration identification	Calibration due date	Verified by:
Pressure gauge	0–100 PSI	35–100 PSI	PM-1651	11/04/23	Technician

in the "Expected Result" column. The "Pass" option was selected for each box. Thus, it was confirmed that the equipment passed the verification. Therefore, it was ensured that the IQs were successful.

4.4.2 Results of Stage 2: Engineering Technical Evaluation

The results of the TEEs are presented at this stage. During the execution of the engineering technical evaluation, a fully factorial DOE, a response surface DOE, and a transmission of variation analysis were performed.

4.4.2.1 Fully Factorial Experimental Design

The DOE was performed during the ultrasonic sealing process. The TEE parameters were evaluated using a full factorial design. The design summary outlines four key factors: sealing time, dwell time, trigger force, and pressure. In addition, 34 runs were required for two replicates.

Table 4.9 shows the data for the full factorial design with a total of four factors. Because the downward velocity does not have a numerical range, the references for this parameter are colored lines that cannot be analyzed in Minitab®.

The average of six samples per run was used to perform DOE analysis. The main effects plot for the seal integrity force (tail) shown in Figure 4.2A indicates that the weld time, trigger force, and pressure interacted and were significant for the process. On the other hand, it can be observed in Figure 4.2B that hold time had no interaction with the other factors. Thus, it was proven that this factor was not significant in the process.

TABLE 4.8 Verification of equipment operation

PLAN				*REPORT*		
Test #	Objective of the test	Expected result	Verification method	Result	Pass/Fail	Verified by:
1	Check the alignment of the sealer, horn and nest.	The horn and nest are correctly aligned with the soldering iron.	Visual	The horn and nest are correctly aligned.	☒ Go to ☐ Failure	Engineer
2	Adjust the mechanical stop of the sealer so that the horn does not touch the nest.	The mechanical stop prevents the horn from contacting the nest.	Visual	The mechanical stop can be adjusted, as expected.	☒ Go to ☐ Failure	Engineer
3	Turn on the sealer.	The power LED indicator will illuminate. The welder will turn on and perform a "Self Check." "READY" will appear on the display indicating completion of the self-check and that the welder is ready for operation.	Visual	The machine is energized correctly.	☒ Go to ☐ Failure	Engineer
4	Adjust the air pressure regulator.	The air pressure is adjusted accordingly.	Visual	The power LED lights up. RDY appears on the display.	☒ Go to ☐ Failure	Engineer
5	Press the DOWN button, followed by the START switches.	The LED on the switch will illuminate. The horn will lower and remain in the down position. Verify that the horn does not touch the nest using the mechanical stop.	Visual	The air pressure is adjusted as expected.	☒ Go to ☐ Failure	Engineer

(*Continued*)

TABLE 4.8 (Continued)

PLAN				*REPORT*		
6	When the horn is in the down position and in the HORN DOWN mode, press the HORN DOWN button again.	The horn will be raised.	Visual	Results as expected.	☒ Go to ☐ Failure	Engineer
7	When the screen lock is inactive, press the Hold Time switch.	It will display the Hold parameter on the NUMERICAL DISPLAY and turn on the Hold Time switch LED.	Visual	The horn rises as expected.	☒ Go to ☐ Failure	Engineer
8	When the screen lock is inactive, press the weld time switch.	It will display the weld parameter on the NUMERICAL DISPLAY and turn on the visual weld time switch LED. Visual	Visual	Results as expected.	☒ Go to ☐ Failure	Engineer
9	When the screen lock is inactive and once you have selected a parameter, press the Up/Down switches.	The values of the Visual parameters should increase or decrease according to the switch pressed.	Visual	Results as expected.	☒ Go to ☐ Failure	Engineer
10	Emergency Stop Functionality	The Emergency Stop button stops the machine operation when pressed. When pressed, the reset button and action buttons should not function.	Visual	Results as expected.	☒ Go to ☐ Failure	Engineer

11	Security Interlock Functionality	The welding operation can only be started by simultaneously pressing both palm buttons until they are fully extended.	Visual	Results as expected.	☒ Go to ☐ Failure	Engineer
12	Weld Cycle Verification	The horn moves to the default position correctly.	Visual	Results as expected.	☒ Go to ☐ Failure	Engineer
13	Reset Switch	Resets any resettable alarm condition (indicated by the switch LED flashing). Pressing and holding RESET prevents a weld cycle from starting. Releasing RESET returns the welder to the READY state, if there are no non-resettable errors. Top of Form	Visual	Results as expected.	☒ Go to ☐ Failure	Engineer

TABLE 4.9 Full factorial design

FACTOR	*UNIT OF MEASURE*	*EQUIPMENT USAGE RANGE FACTOR*
Sealing time	ms	100–250
Waiting time	sec	1–10
Trigger force	Lbs	15–200
Pressure	psi	20–80

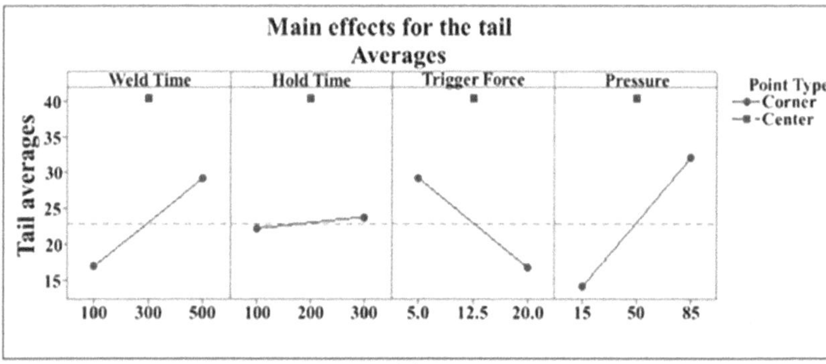

A)

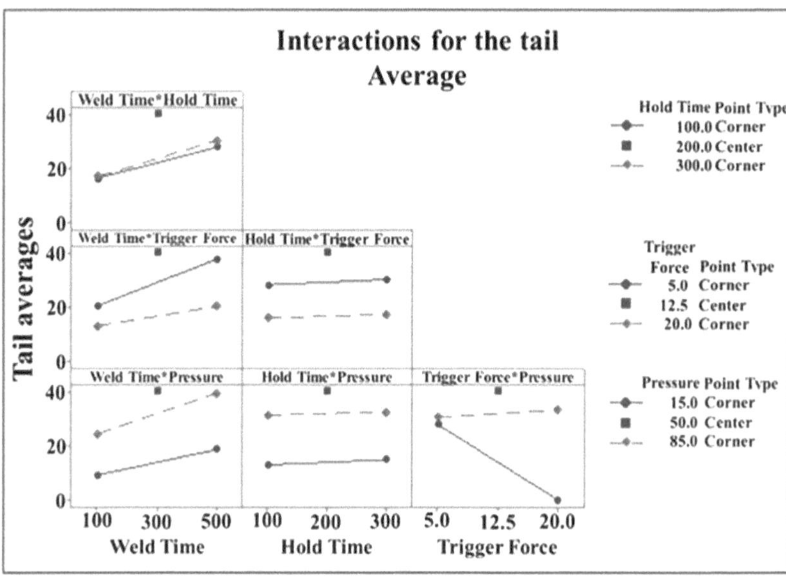

B)

FIGURE 4.2 A) Main effects for the pencil tail, and B) interaction plot of variables.

According to the factorial regression analysis and the results shown in Figures 4.2 and 4.3, it was concluded that the factors (input variables) with a significant effect on the outputs in Table 4.11 are the weld time, pressure, and trigger force. These factors were used to continue the study of the complete process development of the DOE response surface.

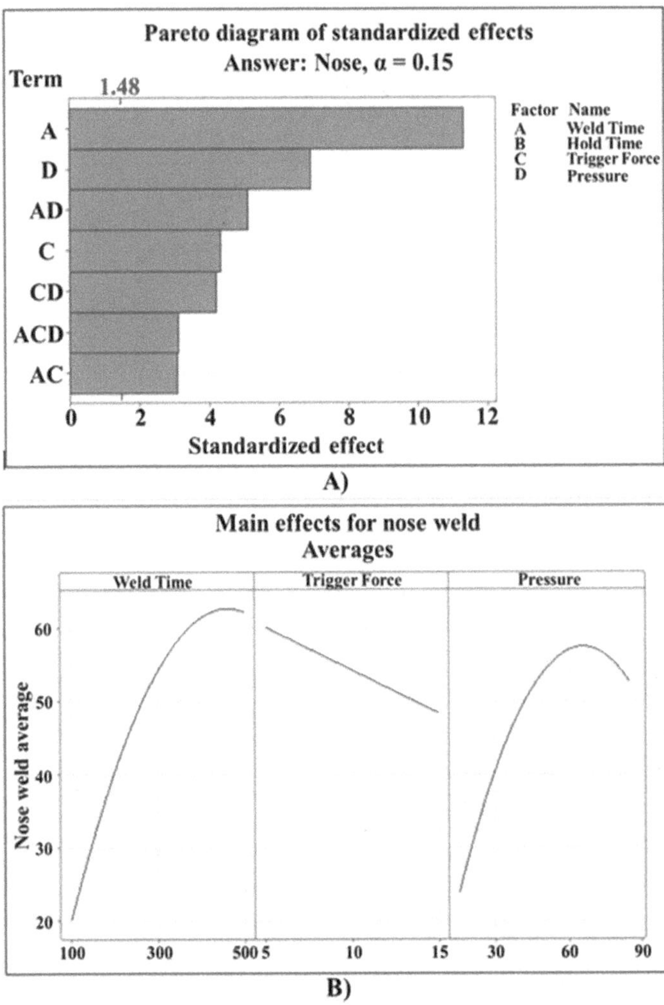

FIGURE 4.3 A) Pareto diagram of the standardized effects and B) main effects for nose sealing

4.4.2.2 DOE Response Surface

According to the DOE fractional factorial design, the input variables with a significant effect on the output were determined. These variables have been identified as the "vital few" controllable factors. The intent of the DOE response surface analysis is to determine the configuration of the factors that optimize the response. For the DOE response surface analysis, a design with 30 runs (15 combinations and two replicates) was obtained using the parameters suggested in Table 4.11, with only the three critical factors considered. Figure 4.4 shows the DOE results.

The contour plots in Figures 4.4C and 4.4D show the natural behavior of the process based on the regression equation, and the results of the manufacturing

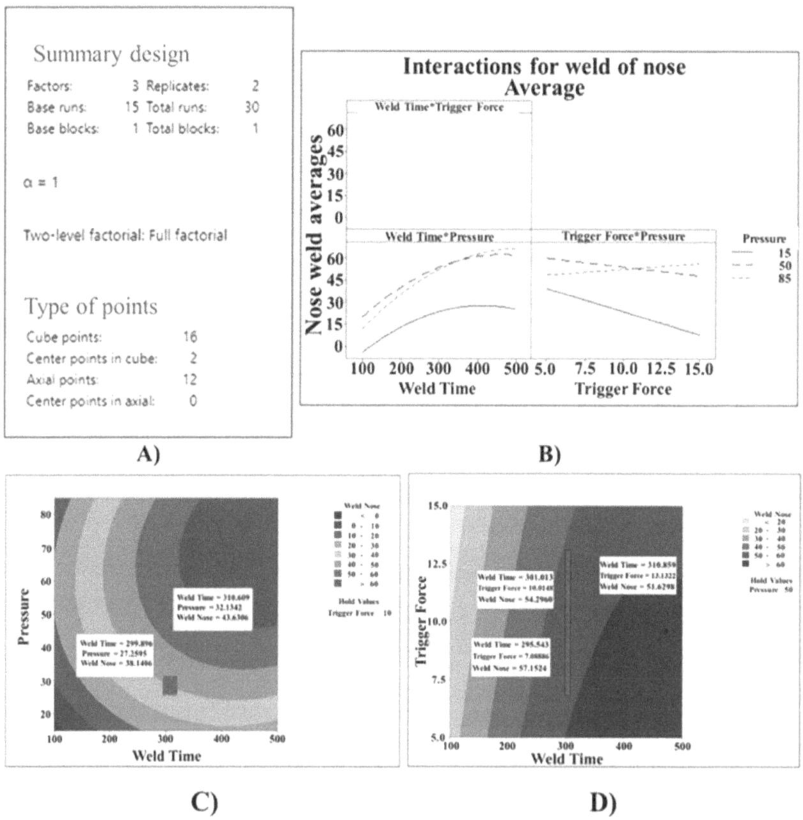

FIGURE 4.4 A) DOE summary, B) interaction for nose weld, C) graph of nose seal contour versus pressure, welding time, and D) graph of nose seal contour versus trigger force, welding time.

run. The parameter window used for optimization in this study is indicated by a red square. The selected parameter window provided a predicted nominal output of 37.88 lbf for the weld nose when the specifications for the weld nose were 6 lbf. Analysis of the response surface design data yielded the summary information shown in Figure 4.5.

The final parameters obtained from the DOE response surface analysis using the optimization graph are listed in Table 4.10.

4.4.2.3 Variance Transmission Analysis

The VTA aims to optimize the average and reduce the variation in the outputs. This is achieved by identifying the key inputs and setting tolerances. Optimization and variation reduction were achieved by setting input targets and tolerances. The main intention of VTA is to reduce the variation in the runs performed and the results. Thirty samples were run in the low pair, and 30 samples were run in the high pair to perform VTA. The material was inspected for Quality Control. Figure 4.6A shows a summary of the analysis model.

After obtaining the torque values through the DOE response surface analysis, these values were entered into the VTA using the regression equation

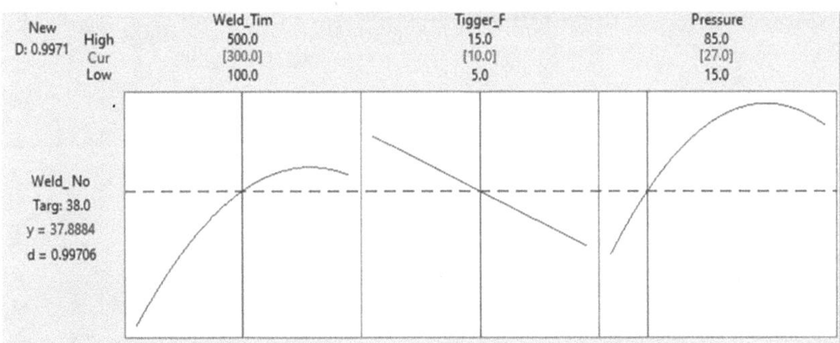

FIGURE 4.5 Analysis of the response surface design data

TABLE 4.10 Normal operating range

PARAMETER	*LOW VALUE*	*HIGH VALUE*
Sealing time	295	305
Trigger force	7	13
Pressure	21	32
Waiting time	200	250

Model summary

S	R-sq	R-sq(adj)	R-sq(pred)
13.4509	83.24%	77.90%	67.63%

Analysis of variance

Source	DF	Adj SS	Adj MS	F-Value	P-Value
Model	7	19763.1	2823.30	15.60	0.000
Linear	3	13632.4	4544.13	25.12	0.000
Weld_Time	1	8822.1	8822.10	48.76	0.000
Tigger_Force	1	689.9	689.92	3.81	0.064
Pressure	1	4120.4	4120.36	22.77	0.000
Square	2	3958.6	1979.29	10.94	0.001
Weld_Time*Weld_Time	1	963.7	963.73	5.33	0.031
Pressure*Pressure	1	1424.2	1424.23	7.87	0.010
2-Way Interaction	2	2172.1	1086.06	6.00	0.008
Weld_Time*Pressure	1	644.9	644.95	3.56	0.072
Tigger_Force*Pressure	1	1527.2	1527.18	8.44	0.008
Error	22	3980.4	180.93		
Lack-of-Fit	7	3848.3	549.75	62.41	0.000
Pure Error	15	132.1	8.81		
Total	29	23743.5			

Regression equation

Weld_Nose = -6.6 + 0.2564 Weld_Time - 3.97 Tigger_Force + 0.882 Pressure
- 0.000328 Weld_Time*Weld_Time - 0.01302 Pressure*Pressure
+ 0.000907 Weld_Time*Pressure + 0.0558 Tigger_Force*Pressure

A)

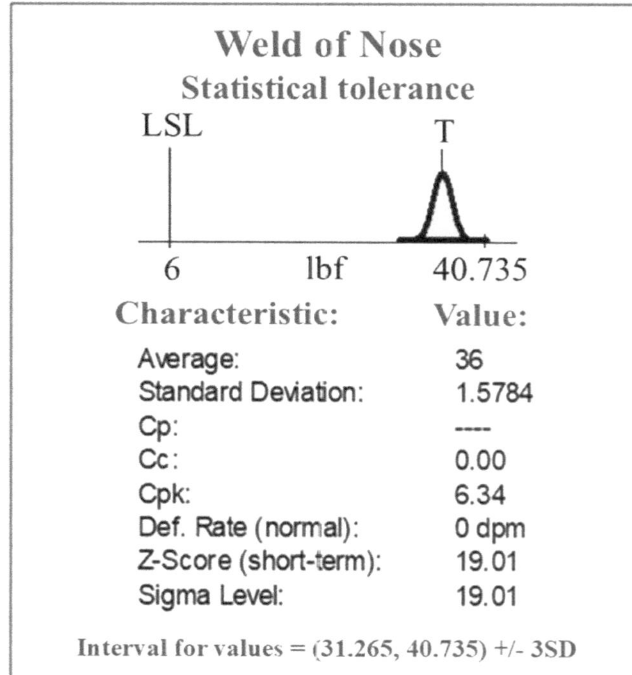

B)

FIGURE 4.6 Results for VTA: A) VTA model summary and B) VTA result

for the weld nose test, which was obtained through Minitab® analysis. This resulted in a Cpk of 6.34 with an average of 36 lbs for the weld nose test, as shown in Figure 4.6B. Figure 4.7 shows the obtained plots. As can be seen, in the normality test plots, except for the one for high tail seal values (Figure 4.7G), a P-value greater than 0.05 was obtained, so the data were normally distributed.

With the results obtained, Table 4.11 is made, where the objectives and tolerances of the inputs are determined, which in this case are the sealing time, trigger force, and pressure. In the interval column for values, the acceptable range in which the equipment can operate according to the factor is placed; this range is given a tolerance of ±3.

Table 4.12 lists the final optimal parameters to be challenged in the OQ stage: weld time, trigger force, and pressure, along with their corresponding values.

4.4.3 Results of Stage 3: OQ of the Equipment

This stage shows the results obtained from the equipment's OQ. Once the TEE study provided the minimum acceptable range, 800 samples within those values were manufactured and used for testing.

Table 4.13 shows the results of the four tests (tests 1, 2, 5, and 6) performed with attribute data, while Table 4.14 shows tests 3 and 4, performed with variable data for the OQ. The "Samples Tested" column shows the sample size for each test performed during the OQ of the equipment. If any of the tests fail, the equipment OQ will not meet the objective, so that it will be rejected. As can be seen, the samples passed all tests.

This validation provides objective evidence that the ultrasonic sealer installed in the fitting-line process produces a weld product that meets predetermined requirements.

The OQ of the ultrasonic sealing process for the Accessory Line and the assembly process exceeded all the conditions set by the protocol. This was accomplished by using the TEE parameter settings developed for the ultrasonic sealer machine's high/low parameters. Therefore, the OQ protocol should be completed for accessory products used in the electrosurgical pencil area.

4.4.4 Results of Step 4: Team Performance Rating

Once the OQ was successfully approved, the sample size of the tests was analyzed. In addition, 1,000 units were manufactured and subjected to various tests for this equipment's PQ.

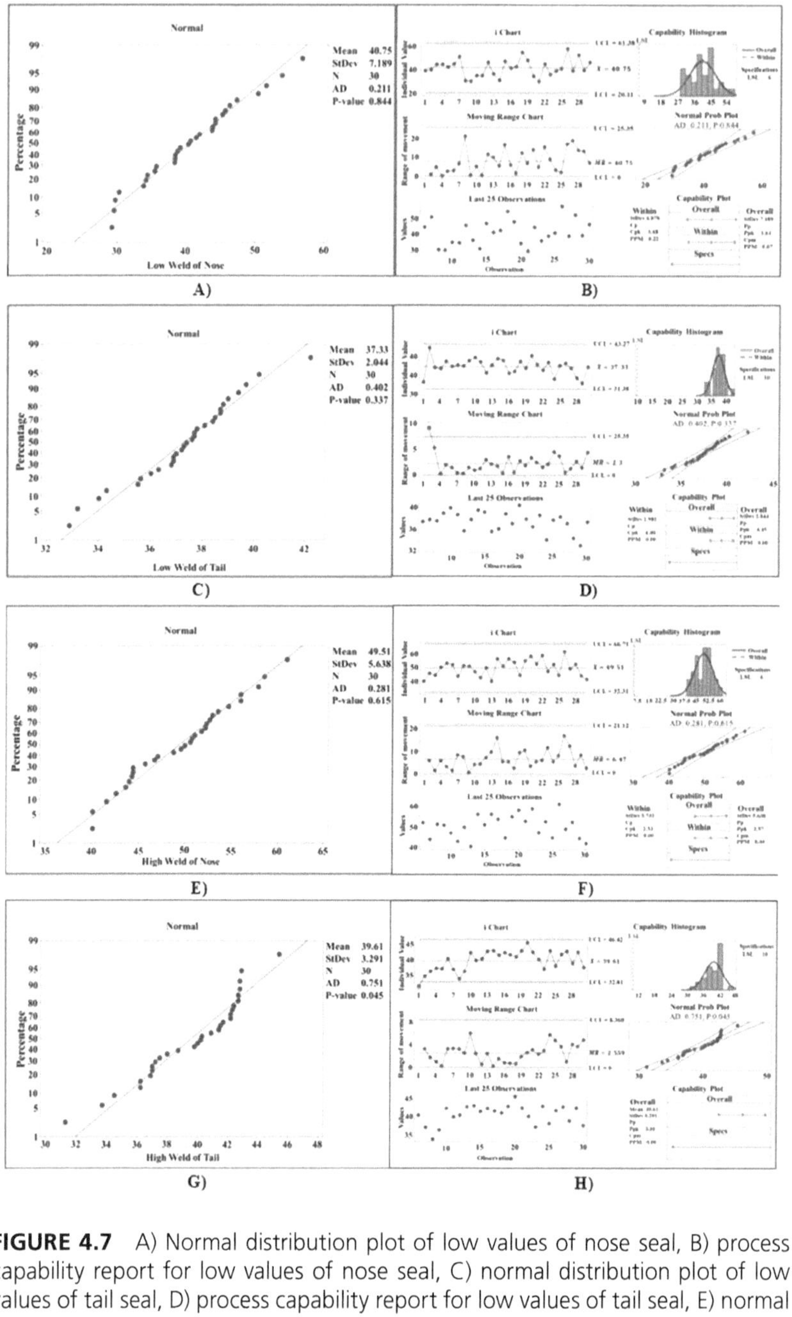

FIGURE 4.7 A) Normal distribution plot of low values of nose seal, B) process capability report for low values of nose seal, C) normal distribution plot of low values of tail seal, D) process capability report for low values of tail seal, E) normal distribution plot of high nose sealing values, F) process capability report for high nose sealing values, G) normal distribution plot of high tail sealing values, and H) process capability report for high tail sealing values.

TABLE 4.11 Input targets and tolerances

ENTRANCE	*TYPE*	*TOLERANCE*	*LSL*	*USL*	*AVERAGE AND VARIANCE REQUIREMENTS*	*RANGE FOR ±3SD VALUES*
Sealing time	Statistician	300 ± 5 ms	295	305	Average = 300, Std. Dev. <= 0.83333	(297.5, 302.5)
Trigger force	Statistician	10.5 ± 3 Lb	7	13	Average = 10.5, Std. Dev. <= 0.5	(9, 12)
Pressure	Statistician	26.433 ± 5.5 psi	21	32	Average = 26.433, Std. Dev. <= 0.91667	(23.683, 29.183)

TABLE 4.12 Tested acceptable range

PARAMETER	*UNDER*	*NOMINAL*	*HIGH*
Sealing time	295	300	305
Trigger force	7	10	13
Pressure	21	26	31

TABLE 4.13 Results of the data analysis by attributes for the OQ of equipment

TEST	*PROCESS OUTPUT*	*PROCESS OUTPUT REQUIREMENTS*	*TESTED SAMPLES*	*NUMBER OF FAILURES*	*DECISION*
1	Visual inspection of surgical pencils	C=0 DC00134072 RE00382896	74	0	☒ Go to ☐ Failure
2	High Voltage Breakdown (Hi-Pot)	Specification: 5.2 KV DC00134072 RE00146810	149	0	☒ Go to ☐ Failure
5	Rocker arm slot width	Test Specification: 0.201″ DC00134072 RE00383105	149	0	☒ Go to ☐ Failure
6	Continuity test and air test cycle of surgical pencils	C=0 DC00134072 RE00385038	74	0	☒ Go to ☐ Failure

TABLE 4.14 Results of variable data analysis

TEST	*PROCESS OUTPUT*	*PROCESS OUTPUT REQUIREMENTS*	*CORRIDA*	*TESTED SAMPLES*	*RESULT*	*DECISION*
3	Weld integrity strength (front)	Test specification: ≥ 6 lbf DC00134072 RE00386523	1	133	P-value < 0.005 PPK = 2.86	☒ Go to ☐ Failure
			2		P-value = 0.954 Cpk = 1.57	
4	Weld integrity strength (back)	Test specification: ≥ 10 lbf DC00070713 RE00386523	1	133	P-value = 0.713 Cpk = 4.20	☒ Go to ☐ Failure
			2		P-value = 0.177 Cpk = 2.34	

TABLE 4.15 Results of data analysis by attribute

TEST	*PROCESS OUTPUT*	*CORRIDA*	*TESTED SAMPLES*	*FAULTS*	*DECISION*
1	Visual inspection of surgical pencils Pass/Fail C = 0	1	113	0	☒ Go to ☐ Failure
		2	113	0	☒ Go to ☐ Failure
		3	113	0	☒ Go to ☐ Failure
2	High voltage break Pass/Fail Specification: 5.2 KV	1	228	0	☒ Go to ☐ Failure
		2	228	0	☒ Go to ☐ Failure
		3	228	0	☒ Go to ☐ Failure
5	Rocker arm slot width Pass/Fail Test Specification: 0.201″ (Gate)	1	228	0	☒ Go to ☐ Failure
		2	228	0	☒ Go to ☐ Failure
		3	228	0	☒ Go to ☐ Failure

TEST	*PROCESS OUTPUT*	*CORRIDA*	*TESTED SAMPLES*	*FAULTS*	*DECISION*
6	*Continuity test and air test cycle of surgical pens Pass/Fail C = 0	1	113	0	☒ Go to ☐ Failure
		2	113	0	☒ Go to ☐ Failure
		3	113	0	☒ Go to ☐ Failure

Table 4.15 shows the results of the four tests (tests 1, 2, 5, and 6) performed with attribute data, while Table 4.16 shows tests 3 and 4 performed with variable data for the PQ. The "Samples Tested" column shows the sample size for each test performed in the equipment PQ. All tests were performed at 99% confidence and 98% conformance levels. If any tests failed, the equipment performance rating would not meet the objective and would be rejected. As can be seen, the samples passed all tests.

This validation provides objective evidence that the ultrasonic sealer installed in the fitting line process produces a weld product that meets the process's predetermined requirements.

The PQ of the equipment for the accessories line and the assembly process exceeded all conditions established by the protocol. The PQ protocol for the equipment used in electrosurgical pencils for accessory products is considered complete.

4.5 CONCLUSION AND RECOMMENDATIONS

4.5.1 Conclusion

The general objective of this project was to validate the ultrasonic sealer equipment using IQ, OQ, and PQ by applying the DOE methodology to meet the company's internal requirements. In addition, two specific objectives were established: (objective) to achieve a cpk greater than or equal to 1.33, and specific objective) to determine the optimal equipment parameters during DOE execution.

TABLE 4.16 Results of the analysis of variable data

TEST	PROCESS OUTPUT	CORRIDA	APPLIED DISTRIBUTION MODEL	P-VALUE	LTL	VCA	DECISION
3	Weld integrity strength (front) Test specification: ≥ 6 lbf	1	Nonparametric	<0.005	53.08	SD = 51.57% SD = 51.57%	☒ Go to ☐ Failure
		2	Nonparametric	<0.005	65.07	SD = 51.57% SD = 51.57%	☒ Go to ☐ Failure
		3	Parametric	0.510	58.82	SD = 51.57% LB for PPK = 1.31	☒ Go to ☐ Failure
4	Weld integrity strength (back) Test specification: ≥ 10 lbf	1	Parametric	0.123	39.72	SD = 49.97 % SD = 49.97 %	☒ Go to ☐ Failure
		2	Nonparametric	<0.005	46.54	SD = 49.97 % SD = 49.97 %	☒ Go to ☐ Failure
		3	Parametric	0.568	42.47	SD = 49.97 % LB for PPK = 1.59	☒ Go to ☐ Failure

LTL = lower tolerance limit, SD = standard deviation, VCA = variance components analysis.

The overall objective was achieved, as all stages of validation were successful. The equipment IQ successfully obtained favorable results in each test. The OQ also obtained a positive result by performing the runs under the parameters determined by the DOE and conducting the corresponding tests as per the quality team's instruction. The result was satisfactory because all the sample units were acceptable, thus complying with specific objective number two. In addition, a Cpk of 6.34 was obtained in the VTA study. This fulfills the first objective of this study. In the PQ, a satisfactory result was also obtained because, when performing the tests required by the quality team for the manufactured units, all units were accepted.

On the other hand, the methodology used in this project was appropriate because the DOE can help determine the optimal parameters of the process variables that impact the quality of a product. Thus, it can be assured that the product will be of high quality. The successful validation of ultrasonic sealers has been a significant achievement in ensuring product quality and process consistency. In addition, by meeting the standards of ultrasonic sealer IQ, OQ, and PQ, it has adequately validated its operation within the established limits, ensuring its functionality for the production of surgical pencils. This validation process demonstrates the equipment's ability to maintain consistent quality in the manufacture of critical medical products, ensuring the final product's safety and providing a solid basis for continuous process improvement and production optimization.

4.5.2 Recommendations

Although the optimal parameters were determined using DOE, periodic evaluations are recommended to check for opportunities to improve the sealing process. Before starting, it is advisable to carry out thorough project planning. In addition to clearly defining the objectives, the necessary resources, the schedule of activities, and the steps to be followed in each phase from validation to implementation. From the beginning of the project, make sure you are aware of all the relevant regulations and standards in the medical industry. This will help ensure that the equipment and processes meet all regulatory requirements.

In addition, it is advisable to keep detailed records of each phase of the project, from conception to completion. This will facilitate review, replication, and process improvements in future projects. Similarly, maintaining the flexibility to adjust the project approach is always a good practice to be prepared in case unexpected challenges arise or opportunities for improvement are identified during development. Adaptability can be the key to success in complex projects.

REFERENCES

Dwivedi, Chandra M. 2006. "Validation of Cell Culture Based Processes and Qualification of Associated Equipment and Facility." In *Cell Culture Technology for Pharmaceutical and Cell-Based Therapies*, edited by Sadettin S. Ozturk and Wei-Shou Hu. 1st ed., 523–52. Boca Raton, FL: CRC Press.

Fields, Timothy J. 2010. "Case Study #2-Questionable Equipment Qualification." *Journal of Validation Technology* 16 (1): 81–83. https://www.proquest.com/openview/e342fe2589b54dc8d5894f6a3127da46/1?pq-origsite=gscholar&cbl=29232.

Jindal, Diksha, Hardeep Kaur, Rajesh Kumari Patil, and Hanumanthrao Chadershekar Patil. 2020. "Validation—In Pharmaceutical Industry: Equipment Validation: A Brief Review." *Adesh University Journal of Medical Sciences & Research* 2 (2): 94–98. https://doi.org/10.25259/AUJMSR_15_2020.

Kaivosoja, Emmi. 2015. *The Role of Usability in Software Validation—Case: Medical Device Manufacturing*. Aalto University. https://aaltodoc.aalto.fi/handle/123456789/18672.

Pluta, Paul L., Timothy J. Fields, and Alan J. Smith. 2010. "Case Study #1-Equipment Cleaning Validation Including Visual Evaluation." *Journal of Validation Technology* 16 (1): 75–80. https://www.proquest.com/openview/aabb3a5147b0d90f436414164900cde6/1?cbl=29232&pq-origsite=gscholar.

Xu, Yuchun, Ashutosh Tiwari, Hao Chen, and Christopher J. Turner. 2018. "Development of a Validation and Qualification Process for the Manufacturing of Medical Devices: A Case Study Based on Cross-Sector Benchmarking." *International Journal of Process Management and Benchmarking* 8 (1): 79–102. https://doi.org/10.1504/IJPMB.2018.088658.